Women in the Chemical Workforce

A WORKSHOP REPORT TO THE CHEMICAL SCIENCES ROUNDTABLE

Chemical Sciences Roundtable

Board on Chemical Sciences and Technology

Commission on Physical Sciences, Mathematics, and Applications

National Research Council

NATIONAL ACADEMY PRESS
Washington, D.C.

NOTICE: The project that is the subject of this report was approved by the Governing Board of the National Research Council, whose members are drawn from the councils of the National Academy of Sciences, the National Academy of Engineering, and the Institute of Medicine.

Support for this project was provided by the National Science Foundation under Grant No. CHE-9630106, the National Institutes of Health under Contract No. N01-OD-4-2139, and the U.S. Department of Energy under Grant No. DE-FG02-95ER14556. Any opinions, findings, conclusions, or recommendations expressed in this material are those of the authors and do not necessarily reflect the views of the sponsors.

International Standard Book Number 0-309-07293-X
Library of Congress Card Number 00-110760

Additional copies of this report are available from:

National Academy Press
2101 Constitution Avenue, NW
Box 285
Washington, DC 20055
800-624-6242
202-334-3313 (in the Washington metropolitan area)
http://www.nap.edu

Board on Chemical Sciences and Technology
2101 Constitution Avenue, NW
NAS 273
Washington, DC 20418
202-334-2156

Printed in the United States of America

THE NATIONAL ACADEMIES

National Academy of Sciences
National Academy of Engineering
Institute of Medicine
National Research Council

The **National Academy of Sciences** is a private, nonprofit, self-perpetuating society of distinguished scholars engaged in scientific and engineering research, dedicated to the furtherance of science and technology and to their use for the general welfare. Upon the authority of the charter granted to it by the Congress in 1863, the Academy has a mandate that requires it to advise the federal government on scientific and technical matters. Dr. Bruce M. Alberts is president of the National Academy of Sciences.

The **National Academy of Engineering** was established in 1964, under the charter of the National Academy of Sciences, as a parallel organization of outstanding engineers. It is autonomous in its administration and in the selection of its members, sharing with the National Academy of Sciences the responsibility for advising the federal government. The National Academy of Engineering also sponsors engineering programs aimed at meeting national needs, encourages education and research, and recognizes the superior achievements of engineers. Dr. William A. Wulf is president of the National Academy of Engineering.

The **Institute of Medicine** was established in 1970 by the National Academy of Sciences to secure the services of eminent members of appropriate professions in the examination of policy matters pertaining to the health of the public. The Institute acts under the responsibility given to the National Academy of Sciences by its congressional charter to be an adviser to the federal government and, upon its own initiative, to identify issues of medical care, research, and education. Dr. Kenneth I. Shine is president of the Institute of Medicine.

The **National Research Council** was organized by the National Academy of Sciences in 1916 to associate the broad community of science and technology with the Academy's purposes of furthering knowledge and advising the federal government. Functioning in accordance with general policies determined by the Academy, the Council has become the principal operating agency of both the National Academy of Sciences and the National Academy of Engineering in providing services to the government, the public, and the scientific and engineering communities. The Council is administered jointly by both Academies and the Institute of Medicine. Dr. Bruce M. Alberts and Dr. William A. Wulf are chairman and vice chairman, respectively, of the National Research Council.

CHEMICAL SCIENCES ROUNDTABLE

RICHARD C. ALKIRE, University of Illinois at Urbana-Champaign, *Chair*
MARION C. THURNAUER, Argonne National Laboratory, *Vice Chair*
ALEXIS T. BELL, University of California, Berkeley
DARYLE H. BUSCH, University of Kansas
MARCETTA Y. DARENSBOURG, Texas A&M University
MICHAEL P. DOYLE, Research Corporation
BRUCE A. FINLAYSON, University of Washington
RICHARD M. GROSS, The Dow Chemical Company
ESIN GULARI, Wayne State University
L. LOUIS HEGEDUS, Elf Atochem North America, Inc.
ANDREW KALDOR, Exxon R&D Laboratories
FLINT LEWIS, American Chemical Society
ROBERT L. LICHTER, The Camille & Henry Dreyfus Foundation, Inc.
MARY L. MANDICH, Bell Laboratories
ROBERT S. MARIANELLI, Office of Science and Technology Policy
TOBIN J. MARKS, Northwestern University
JOE J. MAYHEW, Chemical Manufacturers Association
WILLIAM S. MILLMAN, U.S. Department of Energy
NORINE E. NOONAN, U.S. Environmental Protection Agency
JANET G. OSTERYOUNG, National Science Foundation
NANCY L. PARENTEAU, Organogenesis, Inc.
GARY W. POEHLEIN, National Science Foundation
MICHAEL E. ROGERS, National Institute of General Medical Sciences
HRATCH G. SEMERJIAN, National Institute of Standards and Technology
PETER J. STANG, University of Utah
D. AMY TRAINOR, Zeneca Pharmaceuticals
JEANETTE M. VAN EMON, U.S. Environmental Protection Agency National Exposure Research Laboratory
ISIAH M. WARNER, Louisiana State University

Staff

RUTH MCDIARMID, Senior Program Officer
SUSAN R. MORRISSEY, National Research Council Intern
SYBIL A. PAIGE, Administrative Associate
DOUGLAS J. RABER, Director, Board on Chemical Sciences and Technology

BOARD ON CHEMICAL SCIENCES AND TECHNOLOGY

COMMISSION ON PHYSICAL SCIENCES, MATHEMATICS, AND APPLICATIONS

Preface

The Chemical Sciences Roundtable (CSR) was established in 1997 by the National Research Council (NRC). It provides a science-oriented, apolitical forum for leaders in the chemical sciences to discuss chemically related issues affecting government, industry, and universities. Organized by the NRC's Board on Chemical Sciences and Technology, the CSR aims to strengthen the chemical sciences by fostering communication among the people and organizations—spanning industry, government, universities, and professional associations—involved with the chemical enterprise. The CSR does this primarily by organizing workshops that address issues in chemical science and technology that require national attention.

Women in the chemical workforce was identified as a workshop topic by the CSR in response to broad concerns in the chemical sciences community about underutilization of women in chemistry and chemical engineering. To provide a forum for exploring these concerns, a workshop was planned for May 2000.

The workshop, "Women in the Chemical Workforce," gathered leaders in chemistry and chemical engineering from government, industry, and academia together with experts from the social sciences and from business to explore practical approaches to improving the effectiveness of employers from all sectors in hiring and enabling the success of female professionals in chemistry. I am grateful to those members of the roundtable who provided key assistance in identifying topics and people for the workshop. A special note of thanks goes to Dr. Jong-On Hahm of the National Research Council's Committee on Women in Science and Engineering for her advice and assistance in organizing the workshop.

The papers in this volume are the authors' own versions of their presentations, and the discussion comments were taken from a transcript of the workshop. In accord with the policies of the CSR, the workshop did not attempt to establish any conclusions or recommendations about needs and future directions, focusing instead on issues identified by the speakers.

Janet G. Osteryoung
Workshop Organizer

Acknowledgment of Reviewers

This report has been reviewed in draft form by individuals chosen for their diverse perspectives and technical expertise, in accordance with procedures approved by the NRC's Report Review Committee. The purpose of this independent review is to provide candid and critical comments that will assist the institution in making its published report as sound as possible and to ensure that the report meets institutional standards for objectivity, evidence, and responsiveness to the study charge. The review comments and draft manuscript remain confidential to protect the integrity of the deliberative process. We wish to thank the following individuals for their review of this report:

Nancy B. Jackson, Sandia National Laboratory,
E. Ann Nalley, Cameron University,
Peter W. Rabideau, Iowa State University, and
Geraldine L. Richmond, University of Oregon.

Although the reviewers listed above have provided many constructive comments and suggestions, they were not asked to endorse the conclusions or recommendations nor did they see the final draft of the report before its release. The review of this report was overseen by Marsha I. Lester, University of Pennsylvania, appointed by the Commission on Physical Sciences, Mathematics, and Applications, who was responsible for making certain that an independent examination of this report was carried out in accordance with institutional procedures and that all review comments were carefully considered. Responsibility for the final content of this report rests entirely with the authoring committee and the institution.

Contents

Summary

Janet G. Osteryoung
National Science Foundation

In the fields of science and engineering, women are not represented in proportion to their fraction in the U.S. population. This underrepresentation is especially pronounced in academic departments, where hiring of women lags far behind their representation in the pool of doctoral degree holders. Furthermore, women apparently do not ascend the career ladder as fast or as far as their male counterparts. Recent reports[1,2] elaborate on the impact of this and related issues for science, the academic enterprise, the U.S. economy, and global economic competitiveness. The Chemical Sciences Roundtable judged that the demographics of the workforce and the implications for science and society vary, depending on the field of science or engineering. Accordingly, it organized a workshop, "Women in the Chemical Workforce," to address issues pertinent to the chemical and chemical engineering workforce as a whole, with an emphasis on the advancement of women.

Each of the workshop's three sessions—Context and Overview, Opportunities for Change, and Conditions for Success—included, in addition to presentations by invited speakers, discussion within breakout groups and an oral report from each group.

CONTEXT AND OVERVIEW

The presentation by Margaret W. Rossiter, of Cornell University, was titled "1970-2000: A Less Than Golden Age for Women in Chemistry?" The last 25 years have been a kind of golden age for women in science and engineering in the United States, compared with previous times. Laws were passed in 1972 that, pushed by well-publicized lawsuits, government investigations, voluntary pressure, and individual initiative, made substantial quantitative differences in the training and job opportunities

[1]Office of Science Technology and Policy/National Science and Technology Council, Interagency Working Group, *Ensuring a Strong U.S. Scientific, Technical, and Engineering Workforce in the 21st Century,* Washington, D.C., April 11, 2000.

[2]Commission on the Advancement of Women and Minorities in Science, Engineering, and Technology Development (CAWMSET), *Land of Plenty: Diversity As America's Competitive Edge in Science, Engineering and Technology,* July 13, 2000. Available online at <http://www.nsf.gov/od/cawmset/>.

open to women in chemistry, chemical engineering, and other fields of science and technology. Yet advancement and recognition for women have not occurred as rapidly in chemistry as in biology, despite many notable achievements such as "firsts" by Anna Harrison and a Nobel Prize for Gertrude Elion. One could ask, Why? Is there some unwritten agenda in chemistry (and chemical engineering) regarding career prospects or expectations that needs to be addressed? Are there "best practices" that need to be more widely adopted? Moreover, should incentives be given to those who make improvements?

Why do so few women scientists occupy positions of power and prestige? Professor Virginia Valian, of Hunter College, CUNY, spoke on causes of slow advancement for women in science and engineering and discussed some remedies to speed up that advancement. Although women's underrepresentation at the top of the ladder has many causes, two are particularly evident. Gender schemas—our nonconscious hypotheses about sex differences—are an important cognitive reason that we all, men and women alike, undervalue women's performance and overvalue men's. In addition, small instances of under- or overvaluation accumulate over time to disadvantage women and advantage men.

Arthur Bienenstock, of the Office of Science and Technology Policy, summarized the main findings and ideas of the report *Ensuring a Strong U.S. Scientific, Technical, and Engineering Workforce in the 21st Century.* The nation's economy, knowledge base, and ability to address pressing public health, environmental, and national security challenges in the 21st century will depend greatly on the strength of its scientific, technical, and engineering (ST&E) workforce. Workers in ST&E are essential to both the private and public sectors. Based on a tight global ST&E workforce, changing demographics, and projected growth in ST&E-based jobs, it is in the national interest to pursue the development of domestic ST&E workers from all ethnic and gender groups. We should pay special attention to women, underrepresented minorities, and persons with disabilities. It is these groups, currently underrepresented in the ST&E workforce, that account for much of our nation's growing talent pool.

Breakout Discussions

Following the presentations in the session on context and overview (Chapters 1 through 3), breakout sessions were organized to enable more extensive discussion among the workshop participants. The following questions and statements were suggested to the breakout groups as possible topics for consideration:

1. Tell some stories about your workplace.
2. What gender schemas do you see operating in your workplace?
3. What are the consequences of underrepresentation of women in science and engineering, especially at higher professional levels?

Discussion leaders from the breakout groups then reported in plenary session what they believed to be important ideas and topics brought out in the discussions. The reporters for this session were Sandra C. Greer (University of Maryland), W. Sue Shafer (University of California, San Francisco), Geraldine L. Richmond (University of Oregon), Lou Ann Heimbrook (Lucent Technologies), and Frankie K. Wood-Black (Phillips Petroleum). Stories from the workplace frequently centered on the difficulties experienced by women who find that they are the only woman in a group. The consequences of there being only small numbers of women in highly visible positions also were discussed from the perspective that there are not enough different role models for younger women. A third point of discussion was the need to ensure sufficient numbers of women at every level so that there is a good recruiting pool for positions vacated at higher levels. Much of the discussion elaborated on the notion of schemas discussed by Professor Valian.

She argued that many of the difficulties faced by women in the workplace—difficulties reinforced by the small numbers—arise from discord between the schema for acceptable behavior of women and the schema describing appropriate professional behavior in an organization.

OPPORTUNITIES FOR CHANGE

L. Shannon Davis, of Solutia Inc., offered her perspective on women scientists in industry. Finding women at many levels in the chemical industry is no longer the oddity that it once was. How has the chemical industry seemingly made strides in the hiring, retention, and advancement of women? Some programs and approaches commonly encountered in the chemical industry include formal diversity training, mentoring, and networking. Continuing and improved success depend on executive leadership and the use of specific tactics that have been proven to be effective.

Debra R. Rolison, of the Naval Research Laboratory, spoke on the topic "Title IX for Women in Academic Chemistry: Isn't a Millennium of Affirmative Action for White Men Sufficient?" Chemistry departments need more women as faculty. Yet applications from women for advertised positions are said to approach only about 10 percent of the total, even though one-third of U.S. Ph.D.s in chemistry are awarded to women. Why aren't women chemists applying to academia in proportion to their numbers? Should the logic of Title IX be applied to U.S. chemistry departments? In other words, should federal funds be withheld from those universities that do not increase their faculty hires to reflect the pool of U.S.-granted chemistry Ph.D.s? Can the threat of the loss of federal dollars be the impetus for the changes necessary in U.S. universities to create a departmental environment that women will find hospitable? Many posit that such changes concomitantly will improve the academic experience for all faculty and students. Plausible action items include aggressively recruiting good women candidates for faculty openings, ensuring on-campus day care, mentoring junior faculty in the early stages of their careers, and truly rewarding good teachers and advisors for their skill in guiding and challenging their students. It is not coincidental that these efforts would help men, too.

Breakout Discussions

The following questions and statements were suggested to the breakout groups as possible topics for discussion:

1. What are the negative and positive aspects of formal diversity programs?
2. Compare employment practices in industry and academia.
3. Are the best practices of industry transferable to academia?

Discussion leaders for this session were Frankie Wood-Black, Lou Ann Heimbrook, Geraldine Richmond, W. Sue Shafer, and Maria K. Burka (National Science Foundation). There was considerable discussion about formal diversity programs, with an emphasis on contrasting the situations in industry and academia. Several of the participants concluded that big companies can be successful in recruiting because of stable, well-developed personnel departments and procedures, mandated training, specific metrics used in evaluations of employees, and explicit reward systems. Professor Richmond emphasized the hostile environment often experienced by women faculty as well as the need for women to evaluate their individual situations and establish their own priorities. Professor Shafer emphasized the importance of viewing faculty as an investment. Dr. Burka attributed the weaknesses of academic practices in attracting women faculty to a general lack of accountability on the part of universities.

The workshop organizers and attendees were honored to have as the guest speaker at dinner after the second session the Honorable Eddie Bernice Johnson (D-Texas) of the House of Representatives. Congresswoman Johnson pointed out the importance of science and technology to increasing productivity and our current period of prosperity. These advances are largely the result of past investments and may be inhibited in the future by failure to invest appropriately today. An especially critical concern is the technical workforce, which is decreasing in relative size as demand increases. Furthermore, the pool of potential workers has a demographic composition (e.g., more women, more minorities) quite different from that of the existing technical workforce, which is heavily white and male. It is essential to invest in better education for *all* of our children to realize our country's potential in the 21st century.

CONDITIONS FOR SUCCESS

Kathy E. Sendall, of Petro-Canada, provided a Canadian perspective in her presentation, titled "Gender Diversity in the Workplace: The Leadership and Organizational Imperative."[3] The premise is that women must be attracted to and retained in the technical workforce, with particular emphasis on engineering and the participation of women in the petroleum sector. The business imperative makes it essential that all businesses worldwide embrace a new paradigm in order to remain competitive. The changing needs of industry and emerging expectations of a new engineering and leadership skill set suggest the value of an androgynous leadership style, that is, a style that incorporates the best features of male and female behavioral stereotypes. Data are presented on the Canadian situation, with respect to women in the workplace generally, and in the engineering profession specifically. Leadership by women is critical for going beyond the "programs" in the workplace to effect sustained change.

Nancy H. Hopkins, of MIT, discussed the MIT study on the women faculty in the School of Science, which was prompted by a perception on the part of senior women faculty that they were not treated as well as their male colleagues.[4] The methodologies used in the study included detailed analysis of space allotments, salaries, funds for research, service on committees with power, and so on. The aftermath since the release of the report summary in the spring of 1999 has included widespread publicity, some institutional change at MIT, and rectification of problems in the treatment of women faculty.

Breakout Discussions

The following questions and statements were suggested to the breakout groups as possible topics for discussion:

1. What are the unwritten agendas and folktales of career prospects for women in chemistry and chemical engineering? Are they realistic?
2. Are there rewards for an institution that improves its record with respect to hiring and promoting female chemists and chemical engineers?
3. How rapidly can organizations change? How rapidly must they change?

---

[3]A written contribution for the presentation was not available for inclusion in this report of the workshop.

[4]"A Study of the Status of Women Faculty in Science at MIT," *MIT Faculty Newsletter*, March 1999. Available online at <http://web.mit.edu/fnl/women/women.html>. This article summarizes the findings of a 150-page unpublished report on the same subject prepared in 1994.

The reporters for this session were Frankie Wood-Black, Cecily C. Celby (Radcliffe Institute), W. Sue Shafer, Maria Burka, and Lou Ann Heimbrook. There was considerable discussion of myths that persist—often with negative consequences. Several speakers addressed the notion of rewards for institutions that make progress in hiring and promotion of women. Also discussed in this connection was the concept of a diversity quotient—a quantitative evaluation of an institution's success—that would recognize positive accomplishments and simultaneously identify the failures. Several participants described the importance of women working together and networking. Lou Ann Heimbrook suggested that participants should bring up these issues to the board of directors of the American Chemical Society. Participants were urged to take follow-up action after the workshop, to "do something." As articulated by Frankie Wood-Black, the message was, It's not OK to not try.

1

1970-2000: A Less Than Golden Age for Women in Chemistry?

Margaret W. Rossiter
Cornell University

It is a pleasure to be here with you today, especially since historians rarely get to address chemists, even young ones. By way of introduction I should say that I was briefly (until midway through my sophomore year) a chemistry major, thanks to a very enjoyable course and textbook by Professor Harry Sisler, who made chemistry seem beautiful. But then in sophomore year Professor Louis Fieser made organic chemistry seem monotonous—I could not stay interested in all those ways you can turn an aldehyde into a ketone. The history of science was for me a much better subject, and I have been at it ever since.

While in graduate school at Yale University in the early 1970s, when the women's liberation movement was all around us, I asked the professors in my department (History of Science and Medicine, since discontinued) at one of our Friday afternoon beer parties if there had ever been any women scientists. Certainly none had ever been mentioned in any of our courses. The answer was no, there had never been any. Not even Madame Curie who had won two Nobel Prizes? No, she had been a mere drudge who stirred the pitchblende for her husband's experiments. Such was the state of knowledge (or ignorance or even prejudice) and authority then. Later, after I had completed my degree, I determined to see for myself if there had ever been any women scientists of any sort—and, as they say, the rest is history.[1]

Nowadays you do not have to look far to discover that there have been a lot of women chemists in the past.[2] For our purposes today we can start with Anna Jane Harrison (1912 to 1998) of Mount

[1]Margaret W. Rossiter, "Writing the History of Science," in Jonathan Monroe, ed., *Writing Across the Curriculum* (Ithaca, NY: Cornell University Press, forthcoming).

[2]For starters, try Jane Miller, "Women in Chemistry," in G. Kass-Simon and Patricia Farnes, eds., *Women of Science, Righting the Record* (Bloomington, IN: Indiana University Press, 1990), pp. 300-334, and Maureen M. Julian, "Women in Crystallography," ibid., pp. 335-383; Louise S. Grinstein, Rose K. Rose, and Miriam H. Rafailovich, eds., *Women in Chemistry and Physics: A Biobibliographic Sourcebook* (Westport, CT: Greenwood Press, 1993); and Benjamin F. Shearer and Barbara S. Shearer, eds., *Notable Women in the Physical Sciences: A Biographical Dictionary* (Westport, CT: Greenwood Press, 1997).

Holyoke College, who in 1978 was the first woman elected president of the American Chemical Society. She did this on her third try (having run against Nobelist Glenn Seaborg in her first attempt and a black chemist, Henry Hill, in her second). It had taken over a hundred years (the society was established in 1876) for this "first," although women had been members from the start, or nearly so, and had been heads of sections since Icie Macy Hoobler chaired the Detroit section in 1930.[3] Yet until the mid-1970s there had not been a woman president. Since then there have been others.

By 1978 Harrison was finishing her term on the National Science Board, to which she had been appointed by Richard Nixon in 1972. In the 1980s, she was the fourth woman president of the American Association for the Advancement of Science (the first woman chemist). Thus she was one of the very few women scientists of the 1970s and 1980s to play much of a role in the whole enterprise that is called "science policy." She even ran a meeting on international science at Mount Holyoke (a women's college) in the mid-1980s.[4]

Besides Harrison, and in a sense following in her footsteps, was and is Mary Good (1931-), who has worked in all three sectors of academia, industry, and government. She was both a member and chair of the National Science Board (a possible first for womankind) and was president of the American Chemical Society in 1987. Recently she stepped down as undersecretary in the Commerce Department. She is also rare in having held presidential appointments from three U.S. presidents—Carter, Reagan, and Clinton.[5] Coming along behind her would be Marye Anne Fox, who served in the 1990s in what may be a seat for Southern women on the National Science Board and who is now, after many years at the University of Texas, the chancellor of North Carolina State University.[6] So by now—the year 2000—there have been a few women chemists in top and highly visible and responsible places in the United States.

An internationally renowned achievement for women was the winning of the Nobel Prize in physiology and medicine in 1988 by Gertrude Elion of Glaxo Wellcome in North Carolina.[7] No woman chemist had done this since biochemist Gerty T. Cori, who shared the physiology Nobel with her husband and Bernardo Houssay in 1947. (Marie Curie won the Nobel Prize in chemistry in 1911, her daughter Irene shared it with her husband in 1935, and Englishwoman Dorothy Hodgkin won it in

---

[3] On women in the American Chemical Society, see Margaret W. Rossiter, *Women Scientists in America: Struggles and Strategies to 1940* (1982) (hereinafter *WSA*, vol. 1), pp. 78-79, and Margaret W. Rossiter, *Women Scientists in America: Before Affirmative Action, 1940-1972* (1995) (hereinafter *WSA*, vol. 2), pp. 305, 306-307, 309, 310, and 421n5. The ACS has also employed women, p. 472n43. See also the delightful autobiography of Icie Gertrude Macy Hoobler, *Boundless Horizons: Portrait of a Pioneer Woman Scientist* (Smithtown, NY: Exposition Press, 1982).

[4] Carole B. Shmurak, "Anna Jane Harrison (1912-), Chemist," in Shearer and Shearer, pp. 172-176; also, obituary in *Boston Globe*, August 12, 1998; "UN Panel Hits Technology's Impact on Women," *Chemical & Engineering News* (hereafter *C&EN*) 61(September 26, 1983), p. 7; and Shirley M. Malcom et al., eds., *Science, Technology and Women: A World Perspective*, AAAS Publication 85-14 (Washington, DC: AAAS, 1985). Harrison's papers are at Mount Holyoke College. On Mount Holyoke and other alma maters of women chemists, see Alfred E. Hall, "Baccalaureate Origins of Doctorate Recipients in Chemistry, 1920-1980," *Journal of Chemical Education* 62 (1985), pp. 406-408.

[5] Ellen Horn Stanley, "Mary Lowe Good (1931-), Chemist," in Shearer and Shearer, pp. 148-153; and *American Men and Women of Science* (hereafter *AMWS*), 19th ed. (1995-96), vol. 3, p. 266.

[6] Fox in *AMWS*, 19th ed. (1995-96), vol. 2, p. 1409.

[7] Marilyn McKinley Parrish, "Gertrude Belle Elion (1918-), Biochemist," in Shearer and Shearer, pp. 84-88; and Richard Kent and Brian Huber, "Obituary, Gertrude Belle Elion (1918-99), Pioneer of Drug Discovery," *Nature* 398 (April 1, 1999), p. 380.

1964.[8] Crystallographer Isabella Karle came close when her husband and long-time collaborator shared it with Herbert Hauptman in 1985.[9])

A case could be made that the last 30 years have been a kind of Golden Age for women in science and possibly even for chemistry—golden, that is, when compared to earlier years.[10] There are more women than ever before getting degrees in science and in a wider variety of scientific fields than ever before. Even fields like oceanography, forestry, or nuclear engineering that had once banned women outright now have many women degree holders. Women scientists are also holding a wider variety of jobs than ever before. A few are being promoted to head various organizations and institutions. This is also true of women in general. There are women cabinet secretaries at State, Justice, Health and Human Services, and Labor, two women on the Supreme Court, nine women senators, and several women heads of major foundations. But despite Harrison, Good, and Fox, not many of these other top women have been chemists—yet, anyway. And certainly they are not in proportion to their numbers, as all fields of science have been getting more feminized, from entry levels on up to top honors, although at different rates from each other.

On the other hand, over the last three decades, women chemists have been more notable for their unusually high rate of suicides and their several lawsuits—two signs of serious discontent—than for their successes.[11] So maybe this has not been such a Golden Age for women chemists as it has for some other American women, but more one of silver, or bronze or tin, or lead or some other metal in the periodic table. Various ways to document this mixed situation are documented below.

DIFFERENTIAL RATES OF CHANGE

Since we are assembled here today in the building of the National Academy of Sciences, one way to depict the mixed situation is to consider the number of women elected to the Academy. Between 1923—itself 60 years after the founding of the Academy during the Civil War—and 1970, ten women had been elected, including biochemist and Nobelist Gerty T. Cori. Ten women members in over 100 years is better than one woman president of the ACS, but it is still barely a trickle.[12] After 1970 the situation improved notably, as shown in Table 1.1. Of the 60 persons who are elected each year, there

[8] Miriam Rossi, "Dorothy Crowfoot Hodgkin (1910-1994), Crystallographer," in Shearer and Shearer, pp. 181-186; and Georgina Ferry, *Dorothy Hodgkin: A Life* (New York: Granta Books, 1998). On Lise Meitner, a famous near-Nobelist, see Clara C. Callahan, "Lise Meitner (1878-1968), Physicist," in Shearer and Shearer, pp. 263-268; and Ruth Sime, *Lise Meitner, A Life in Physics* (Berkeley: University of California Press, 1996).

[9] Janet Owens, "Isabella L. Karle (1921-), Crystallographer," in Shearer and Shearer, pp. 217-222. The Karles' daughter, Louise Karle Hanson, a chemist at Brookhaven National Laboratory, regretted that her mother, who had shared her father's work for so many years, had not been included in the award (John Noble Wilford, "Jerome Karle," *New York Times,* October 17, 1985, p. 17).

[10] Historians apply the term "golden age" retrospectively to a period of great accomplishment, creativity, and opportunity. If alive during such a stretch, one is generally too busy to use the term for a situation that seems normal, but as the good times wane for whatever reason—the funding runs out, the institute closes, the great mentor dies—some begin to perceive a sense of ending and closure, somewhat like a closed parenthesis, that needs a suitable name.

[11] "Women Chemists Mortality Study Finds High Suicide Rate," *C&EN* 62 (April 23, 1984), pp. 16-17. See also Molly Gleiser, "Suicide Among Women Chemists," *Nature* 328 (July 2, 1987), p. 10. She tells her own vivid tale of an unhappy career at the Lawrence Berkeley National Laboratory in Molly Gleiser, "The Glass Wall," in Anna Pattatucci, ed., *Women in Science, Meeting Career Challenges* (Thousand Oaks, CA: Sage Publications, 1998), pp. 204-218.

[12] Rossiter, *WSA*, vol. 1, pp. 285-288, and vol. 2, pp. 326-327.

TABLE 1.1 Women Elected to the National Academy of Sciences

Year	Biological Sciences	Physical Sciences	Social Sciences	Math/Statistics
1970	2	0	0	0
1971	1	0	1	0
1972	1	1	0	0
1973	3	0	0	0
1974	2	0	0	0
1975	1	1	3	1
1976	1	0	0	1
1977	3	0	1	0
1978	2	2	1	0
1979	3	0	0	0
1980	1	0	0	0
1981	2	1	0	0
1982	1	0	0	0
1983	4	0	0	0
1984	4	0	0	0
1985	2	2	0	0
1986	1	1	0	0
1987	0	0	1	1
1988	2	0	2	0
1989	2	0	2	0
1990	3	1	1	1
1991	3	1	1	0
1992	1	3	1	0
1993	3	1	1	1
1994	4	3	1	0
1995	3	1	1	0
1996	7	1	2	1
1997	3	3	1	0
1998	6	0	1	1
1999	5	2	1	1
2000	3	1	2	2

SOURCE: M. Rossiter; data taken from National Academy of Sciences membership lists.

has been at least 1 woman chosen each year and as many as 11 in one year. But what is of interest here is the distribution of scientific disciplines of these women as well as their number. Of the 138 women elected since 1970, 79, or by far the majority, have been in biological fields, including biochemistry. (But even there, Gertrude B. Elion was not elected until 1990, 2 years after sharing the Nobel Prize.) Only 25 of the women have been in the physical sciences, and only 6 of these in chemistry or crystallography: Isabella L. Karle, JoAnne Stubbe, Alexandra Navrotsky, Marye Anne Fox, Judith P. Klinman, and Sylvia T. Ceyer.[13]

[13]Some, but astonishingly not all of these, are in *AMWS*.

TABLE 1.2 Women Elected to the National Academy of Engineering

	Decade			
Section	1970	1980	1990	2000
Aerospace	0	2	0	0
Bioengineering	0	0	0	1
Chemical engineering	0	0	2	1
Civil engineering	0	0	1	0
Computer science	2	3	5	1
Electric power/energy systems engineering	0	0	1	0
Electronics	2	3	2	1
Operating systems	0	0	4	0
Materials	0	3	5	0
Mechanical engineering	0	0	3	0
Petroleum, mining, geological	0	0	2	1
Interdisciplinary	2	3	3	0

SOURCE: M. Rossiter; data taken from National Academy of Engineering membership lists.

In the 1970s there appears to have been much making up for past omissions: in Academy elections many of the women elected were well along in years, often at or past retirement. Lately, however, they have been younger—even in their forties—closer in age to the men being chosen, a sign that they are realistic candidates in their prime years. Election to the NAS in their still-active years also makes the women more powerful figures on their campuses and across the nation.

For the sake of the chemical engineers present, we should next consider the National Academy of Engineering (Table 1.2), which was not established until 1965 but currently has three women (out of 164, or 1.8 percent) in its chemical engineering section—Elisabeth M. Drake, Johanna Levelt Sengers, and Shirley E. Schwartz. In addition, there are several women chemists and chemical engineers who are (or were) members of other engineering sections, such as Edith Flanigen, Mary Good, Elsa Reichmanis, Della Roy, Maxine Savitz, Kathleen Taylor, Nancy Fitzroy, and the deceased Judith Schwan.[14]

Thus what might be called the premier "reward structure" for American scientists and engineers, academic or otherwise, does not seem to be working very well for women chemists and chemical engineers. Of the many thousands who have earned degrees in chemistry and chemical engineering, hardly any are being elected to the top honors in the land.

Since 1970, there has been a tremendous rise in the numbers and percentages of women earning degrees, including Ph.D.s, in chemistry in the United States. Much of these data for the early years were collected by Betty Vetter, who died a few years ago. She was trained as a chemist (with a master's degree from Stanford in the late 1940s) but made a career out of the new field of science manpower statistics. She worked for over 30 years at what is now called the Commission on Professionals in Science and Technology,[15] some of whose voluminous data are presented in Figures 1.1, 1.2, 1.3, and

[14]Rossiter, *WSA*, vol. 2, p. 326.

[15]Vetter in *AMWS*, 19th ed. (1995-96), vol. 7, p. 399. American chemists have been studied more extensively and quantitatively than scientists in other fields, because the ACS has conducted frequent surveys of salaries, degrees, and other factors.

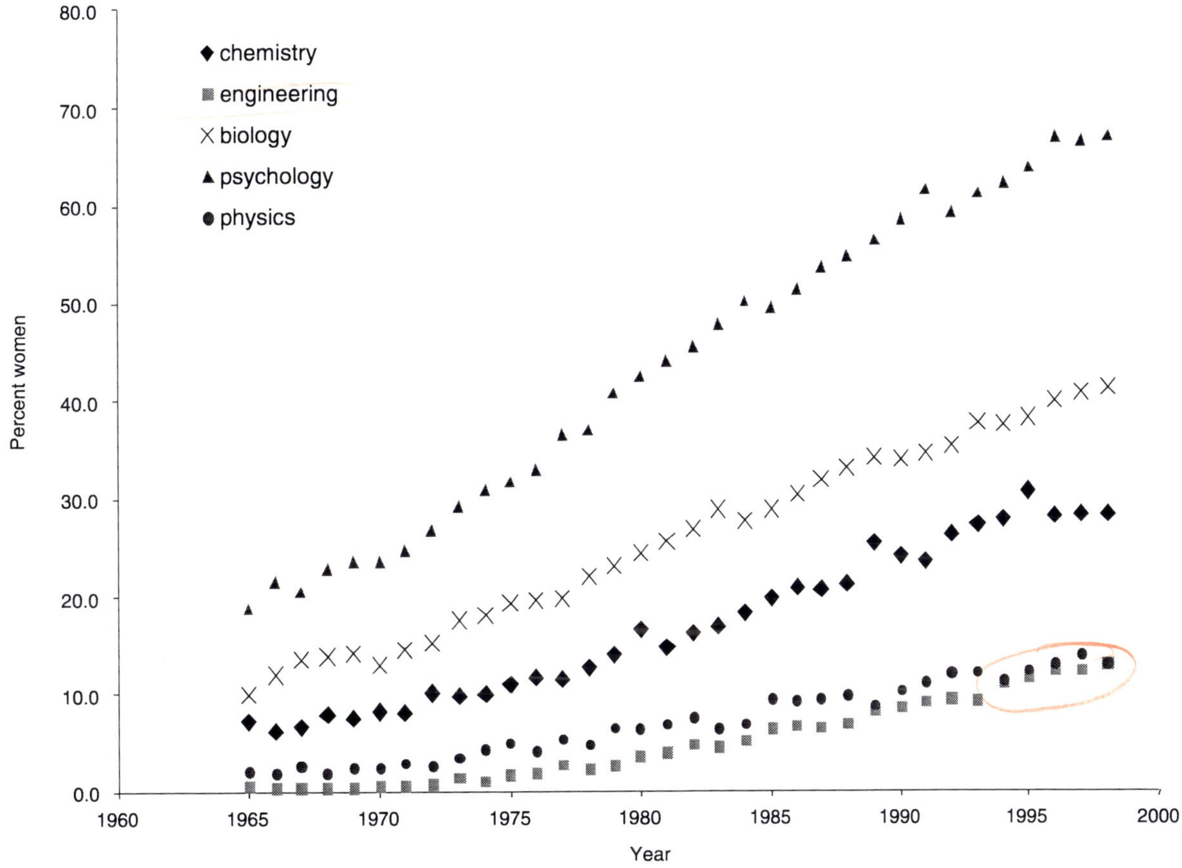

FIGURE 1.1 Women Ph.D.s in science. SOURCE: Commission on Scientific and Technical Personnel Tables, 2-16 and 6-1; data derived from National Science Foundation, *Science and Engineering Degrees, 1966-96,* and *Science and Engineering Doctorate Awards: 1998.*

1.4. Figure 1.1 shows the rising proportions of Ph.D.s going to women in several areas of science and engineering since the 1960s. It is quite remarkable. One thing to notice is that overall the fields that were the most feminized in 1970 or so still are. Psychology and biochemistry (not shown) tolerated 14 to 18 percent women Ph.D.s back then and still have the highest proportions today. Near the bottom of the chart there have been some changes in rank order. Engineering was and still is at the bottom, but agricultural and earth sciences (not shown) have moved up a bit as doors have opened to women in those fields. As for women in chemistry, they had by the early 1990s moved up to 25 percent of the Ph.D.s, and today (2000) the proportion is over 30 percent. One final consequence to note from this chart is a generational change. Someone who got his or her doctorate in 1970 or 1980, who might now be a full professor or in management and who persists in thinking the world of graduate students is still pretty much like it was back then, can be seriously out of touch with current realities. This is important because it is astonishing what a large role autobiography plays in management and personnel matters.

Figures 1.2 and 1.3 provide more detailed data, by year and by subfield, on chemical and engineering degrees. The numbers of women earning doctorates in chemistry over the last three decades was growing not only because of a rise in the number of women entering the field but also because of a drop

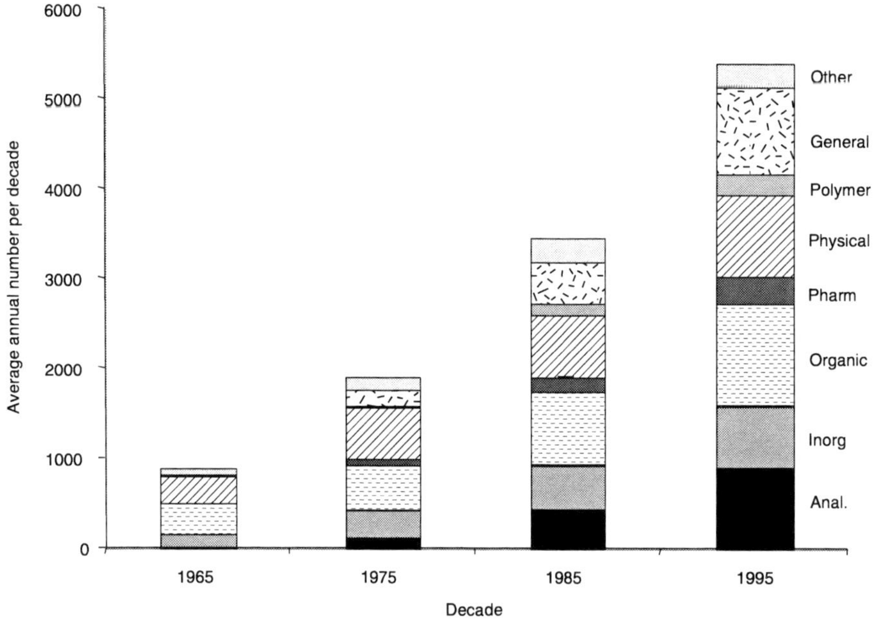

FIGURE 1.2 Women chemists. NOTE: 1995 "decade" includes only 1990 to 1998. "Other" includes agricultural and food chemistry; "physical" includes nuclear and theoretical chemistry. SOURCE: Commission on Scientific and Technical Personnel, Table 6-11; data derived from National Science Foundation, *Survey of Earned Doctorates 1960-1998.*

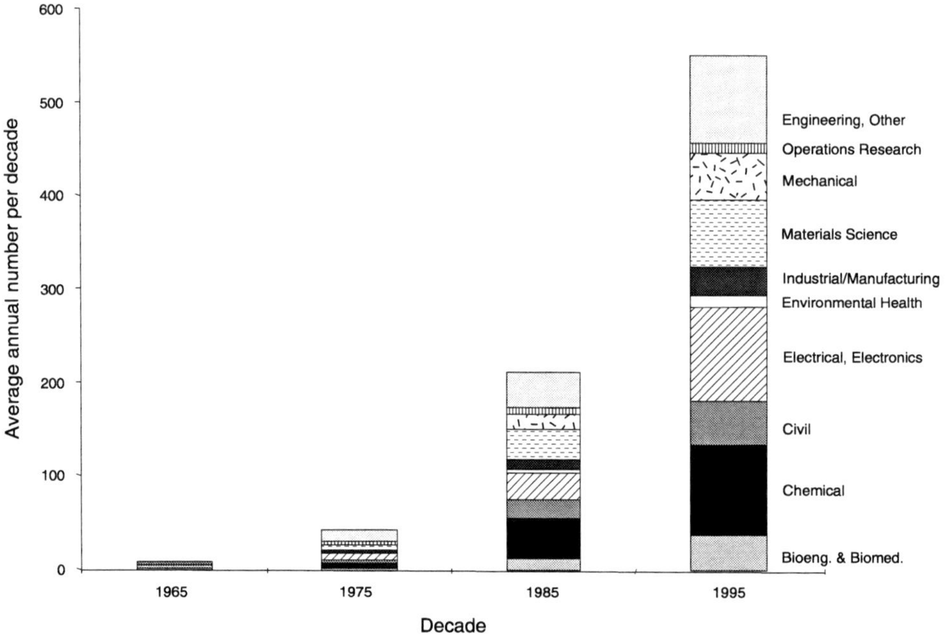

FIGURE 1.3 Women engineers. SOURCE: Commission on Scientific and Technical Personnel, Table 7-1; data derived from National Science Foundation, *Survey of Earned Doctorates, 1998.*

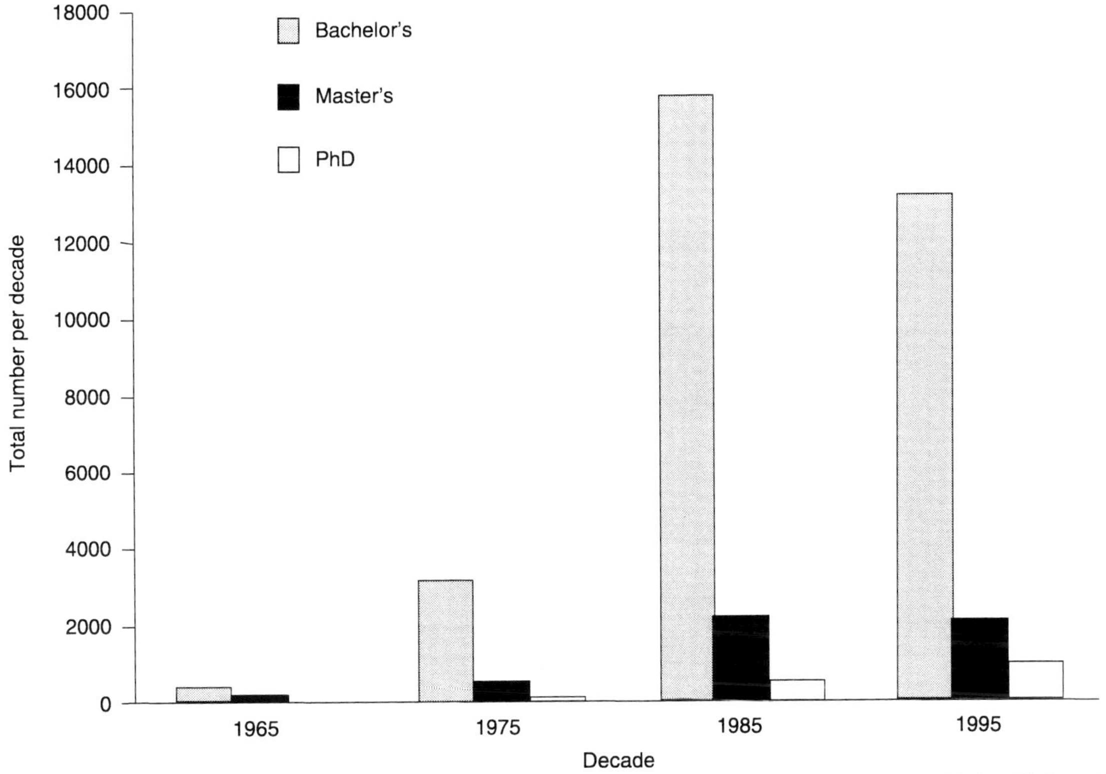

FIGURE 1.4 Women chemical engineers. NOTE: 1990 "decade" includes only 1990 to 1998 for Ph.D.s, 1990 to 1997 for B.S. and M.S. degrees. SOURCE: Commission on Scientific and Technical Personnel, Tables 7-1, 7-04, and 7-05; data derived from National Science Foundation, *Survey of Earned Doctorates, 1998.*

in the number of men majoring in chemistry. The total number of Ph.Ds being awarded to men and women in the physical sciences dropped about one-third in the late 1970s relative to the early 1970s and then rebounded. Thus the total number given now is about the same as it was back in 1970. This should perhaps be an issue of concern to leaders of the field. One might expect there to be some resentments when women become such a rapidly increasing percentage of a pie that is not growing. Chemical engineering has by contrast been growing rapidly at the doctoral level. The numbers of women earning degrees really took off in the 1980s, as shown in Figure 1.4. Currently women are earning about 100 doctorates per year in chemical engineering. Because many chemical engineers still do not need doctorates for full and interesting careers, data are also added here on master's degrees awarded in chemical engineering.

CHANGEMAKERS

If results were correlated with effort, women chemists would have changed academia a lot more by now than has been the case. Although important legislation was passed in 1972 affecting education and employment, a law is only as strong as its enforcement, and fighting powerful institutions in court can be a slow, expensive, and painful process. Nevertheless some women chemists have been quite active in this area. Recall Sharon Johnson, who earned a Ph.D. in chemistry from MIT, was an assistant professor of biochemistry in the medical school at the University of Pittsburgh, and sued her employer in 1973 after her department denied her tenure. The case attracted attention at its outset, when a federal judge issued an injunction prohibiting the university from letting her go before her research grant ran out. This frightened the university's lawyers as well as administrators across the nation, who had no formal evaluation or grievance procedures with which to document their often arbitrary decisions and who were used to having the few personnel matters that led to lawsuits routinely thrown out of court. This pioneering case lasted from 1973 to 1977 because the university, which had superior financial resources, delayed and delayed. (The University of Pittsburgh used at least three and perhaps as many as ten lawyers and spent from $300,000 to $1,000,000 on its defense over 5 years. Johnson and her lawyer, Sylvia Roberts, supported by a loan from the American Chemical Society and donations from the National Organization for Women and the then-new Association for Women in Science (AWIS), persisted through it all, spending under $30,000.) Eventually, in 1977, Johnson lost not on the merits of her case but because the judge finally decided not to intervene in an academic matter. She later joined NIH as a grants administrator, a position that reportedly frightened many academic chemists as it might give her the chance to get revenge by denying them grants.[16]

Also in the early 1970s, another angry chemist, Shyamala Rajendar, a 1965 Ph.D. from the University of Wyoming, was told that her assistant professorship at the University of Minnesota (which she had finally obtained in 1969 after several other staff positions, including research associate) was not on the tenure track after all. (That distinction was introduced about then; before then, all assistant professorships were assumed to be what is now called tenure track.) She sued, sought, and got class-action status when she showed the pattern was industrywide, and finally in 1980 the university settled out of court with a special master appointed to introduce new practices and oversee some rectification of claims.[17]

One can't help but speculate that Rajendar's consciousness grew in her years in the remarkable chemistry department at Wyoming, where Sara Jane Rhoads and N. Rebecca Raulins were on the faculty. Rhoads had come to Wyoming in the late 1940s fresh from Columbia University graduate school and was lucky to get any faculty position in a chemistry department at a coeducational university. She devoted her career to building up the department at Wyoming, was one of only two women chemists among the 163 in the entire country to be awarded an NSF senior postdoctoral fellowship between 1956 and 1971,[18] and won a few other national awards, including the Garvan Medal of the ACS in 1982.[19] Quite aware of inequities in the profession, she may have encouraged young Rajendar to take advantage of the new laws and to fight back. Later on she commented favorably on Rajendar's presentation in 1971 to the Women's Committee of the ACS of the first list of "zeroes," or data compiled from the

---

[16]Johnson in *AMWS*, 12th ed. (1972), vol. 3, p. 3076; "Controversy Heats Up over Tenure Policies," *C&EN* 51 (April 2, 1973), p. 9; "Injunction Granted in Sexist Tenure Case," *C&EN* 51 (June 11, 1973), pp. 2-3; Rebecca L. Rawls, "Female Biochemist Loses Tenure Case," *C&EN* (August 15, 1977), p. 22; Constance Holden, "Court Rules Against Woman Biochemist," *Science* 197 (19 August 1977), p. 743. Her papers on the case are at the Schlesinger Library at Harvard University.

[17]Rajendar in *AMWS*, 12th ed. (1972), vol. 5, p. 5088. See also "Notice of Sex Discrimination Claim Procedures, University of Minnesota Class Action Suit," *Science* 210 (November 28, 1980), p. 1055. She left Minnesota, became a patent attorney in San Francisco, and is now retired and living in Danville, California. She might merit a Garvan Medal. (Judith Ann Trolander, "The Effects of Gender Discrimination Litigation on Academia: The Rajendar Consent Decree," paper presented at the annual meeting of the Organization of American Historians, Toronto, April 1999.)

[18]Rossiter, *WSA*, vol. 2, pp. 318-319. The other was Darleane Hoffman.

[19]Rhoads in *AMWS*, 12th ed. (1972), vol. 5, p. 5206; and "An Interview with Dr. Rhoads and Dr. Raulins," *Wyoming Quarterly* 1, no. 4 (Spring 1976), copy in Sara Rhoads Papers, box 1, folder 3, American Heritage Center, University of Wyoming. Raulins was an instructor at Wyoming who earned her doctorate there in 1953 and was then promoted to the faculty (*AMWS*, 12th ed. (1972), vol. 5, p. 5124).

TABLE 1.3 Number of Women on University Chemistry Faculties in the Academic Year 1969-1970

Number of Women	Number of Faculties
0	113
1	41
2	13
3	2
4	0
5	2
6	0
7	1

SOURCE: *Chemical & Engineering News,* May 10, 1971, p. 21.

ACS's list of faculty at graduate departments, where the University of Wyoming ranked relatively high with its two female faculty members.[20]

The list turned out to be a useful and long-lasting consciousness-raising device. Sister Agnes Ann Green, I.H.M., longtime chair of the chemistry department at Immaculate Heart College in Los Angeles, updated it annually until the mid-1980s, when others took over.[21] A copy of the 1970 version is presented in Table 1.3. It shows that 113 departments had no women chemistry faculty at all. This included Bryn Mawr, Cal Tech (Jacqueline Barton was still an undergraduate at Barnard College),[22] Illinois, MIT, and Minnesota (despite Rajendar's title). The only woman at any of the nine campuses of the University of California system was Marjorie Caserio at Irvine.[23] The University of California at Berkeley hired its first woman in 1978, and UCLA tenured its first woman (Joan Valentine) in 1980. Jean'ne Shreeve was the lone woman at the University of Idaho, and Madeleine Joullié was the one at the University of Pennsylvania.[24] The one woman at the University of Southern California might have been adjunct professor Marjorie Vold, the only woman chemist to get a Guggenheim between 1940 and 1970.[25] By contrast, both Brooklyn Polytech and the newly combined Douglass-Rutgers department had four women, and Rockefeller and American Universities each listed three—all relatively high numbers that put them ahead of Wyoming.

Since 1970 the number of graduate chemistry departments with no women faculty has dropped to fewer than ten institutions. Some, especially private and Roman Catholic ones, have been helped in the last decade with grants from the Clare Booth Luce Foundation that was established in the mid-1980s.

[20]"Few Women in Academia," *C&EN* 49 (May 10, 1971), p. 21.

[21]Green, in *AMWS*, 12th ed. (1972), vol. 2, p. 2288.

[22]Barton, in *AMWS*, 19th ed. (1995-1996), vol. 1, p. 422.

[23]Nancy Allee, "Marjorie Caserio (1929-)," in Shearer and Shearer, pp. 46-51.

[24]Cassandra S. Gissendanner, "Jean'ne Marie Shreeve (1933-), Chemist," in Shearer and Shearer, pp 367-373; and Nina Matheny Roscher, "Madeleine M. Joullié (1927-), Chemist," in Shearer and Shearer, pp. 212-222.

[25]Sharon Sue Kleinman, "Marjorie Jean Young Vold (1913-1991), Colloid Chemist," in Shearer and Shearer, pp. 401-404.

But even if most chemistry departments now have their first woman faculty member—a goal that must have seemed utopian in the early 1970s—this is just the threshold. There is no power in being the only woman, especially if only an assistant professor, who may be temporary; but it can be the first step to more later on. It is just a coincidence but it seems prophetic that academic women chemists have arrived on this threshold at the start of a new millennium. How to bring about more change in the next decade and beyond is what this conference is about. Who should care more about this than women chemists themselves?

CONCLUSION

We live in interesting times. The proportion of women in many areas of American life—including law, politics, medicine, and science—is increasing, dramatically in some cases. But many fields of science are doing better than chemistry—including many but not all of the social sciences and biology. Chemistry departments are just now getting to the point where almost all have one or more women faculty members. (This may not be the case in chemical engineering departments.) This progress has taken 30 years and still provides just the threshold. When one considers that women have been a rapidly rising proportion of the Ph.D.s awarded in the field for 30 years, there ought to have been more change, but perhaps getting from zero to one is the hardest step.

It is time to visualize a future with more women chemists and chemical engineers visibly running things. Not everyone is an Anna Harrison or a Mary Good, but the chemical enterprise is vast and offers challenging tasks at every level. In particular, management is challenging, could be improved at most places, and is well-paying, as every salary survey shows. Maybe in 20 years or preferably sooner, the Chemical Sciences Roundtable will run another conference on the stresses and strains faced by women holding prominent and powerful positions—as presidents of organizations, full professors at major universities, department chairs, deans, and major prizewinners, including more members of the National Academies. Maybe such women will even be using all that visibility and power to make this world a better place.

EPILOGUE

After attending the meeting at which an earlier version of the above was delivered, I am heartened by the energy and commitment of the women chemists and chemical engineers present. Several have formed a group (Committee on Women in Academic Chemistry, or COACh) to develop their management skills and to push and be ready for bigger opportunities in the future. This is exciting indeed.

DISCUSSION

Marcetta Darensbourg, Texas A&M: Did the biologists go through such a conscious-raising exercise as the chemists are trying to go through now? Were they in our position at one time and their numbers grew in job positions, or have they always been a part of the workforce?

Margaret Rossiter: I think they were angry and feeling excluded about 1970, and they met in little clusters, maybe in laboratories, maybe at meetings, and began to figure out what levers ought to be pushed and what resolutions ought to be passed and began to network.

Elizabeth C. Theil, Children's Hospital, Oakland Research Institute: I am a biochemist who was at North Carolina State University for many years. There were always women in biology in lower-level positions. They have always been technicians and lab managers for men, and I think what happened in the 1970s was that there was an awareness that you didn't have to do that anymore. There were the numbers at that point so that they could talk to each other.

Yes, they did get together in groups, but one of the reasons they were able to do it was because they had always been there, just in different roles. That is not the case in chemistry.

Rosemarie Szostak, Army Environmental Policy Institute: I want to make a comment concerning the legal aspects of this, having gone through the lawsuit route myself. If you have one woman [in a department], legally you are not discriminating. So, I think we may be seeing fewer lawsuits as a way of solving the problem of getting the populations of women in the positions at universities. From the standpoint of having gone through the process—and yes, it takes 4 to 5 years—if the other side, the university, can show that they have one woman in any position, then you cannot show discrimination. That is kind of a legal glitch we have to deal with.

Marylee Southard, University of Kansas, Chemical Engineering: What percent of women getting Ph.D.s went to industry and decided not to fight the madness of the academic environment?

Margaret Rossiter: I have no idea. There are probably numbers on who goes into industry, and I think I saw a handout. I think our next speaker is going to talk at length about industry, but then I am not sure industry is a golden place either.

Nina M. Roscher, American University: I have looked at some of these statistics, and one of the things that looking at the faculty at universities doesn't tell you is that many of the women chemists were in universities but were not in faculty positions. It is only in the 1980s that the number of women going into industry started to correspond percentagewise to men. Back in the 1970s, probably about 55 percent of the women were in academia—but they were in postdoc positions, nonteaching, or no-rank teaching positions. So that is where they were, not industry.

Margaret Rossiter: I always used to wonder. They are like lemmings, aren't they? They head to academia. They know they are treated badly, but somehow they are still going there, and somehow it took a while for alternatives to come to their attention.

Mary Ellen Murphy, St. Joseph College: I think one of the problems in academia is the fact that department chairs and deans and vice presidents and presidents very often were men, and so the woman probably didn't have much of a chance or much of a voice at the decision-making level. I think it is only more recently that you have somewhat honest search committees, if most places could even say that at this point in time. Having been a dean, I know very well that there is a tremendous prejudice—even on the part of women—that the man is probably going to be better in the stack of applications. So I think that we have to be concerned about trying to get the jobs, but I also think we have to be concerned about having women in some of these higher-level positions where the decisions are finally made.

Margaret Rossiter: There was a stage when we had the token woman administrator, and I think she was under a lot of pressure and probably couldn't step forward and do too many drastically different things. She probably had a complicated calculus of what she could do and still hold onto her position. I think some of them had fairly short tenure anyway.

Sally Chapman, Barnard College: Somebody will probably be showing these data at some point later in this workshop, but I think that it is very interesting to break down chemistry the same way you have broken down science. There are huge disparities in the advancement of women based on what subdivision of chemistry they are in. The ACS Directory of Graduate Research is computer searchable, so you can assemble these numbers very easily compared to the heroic efforts that were required decades ago. If you look at women in organic chemistry, women in physical chemistry, etc., there are fascinating and quite striking differences.

Margaret Rossiter: How do you explain that?

Sally Chapman: I think there are all sorts of interesting hypotheses. I think organic chemistry is far and away the worst; the number of women in academia who advance to high places in organic chemistry probably can be counted on the fingers of one hand.

Physical chemistry, surprisingly, is relatively good. By "surprisingly," I mean that I am a physical chemist—and when I was a student, I had not expected to see this. When I was in graduate school in the early 1970s, it felt as if we were part of a leading wave of women entering chemistry, and we thought about how this ocean of women would advance. At that point I naively figured that there was a spectrum—from biology, where there were more women, toward physics, where there were fewer. I expected that women in chemistry would be distributed along that particular axis, so physical chemistry would be an area where women did not have as many opportunities.

Physical chemistry is actually pretty good. Now, my hypothesis says (and other people can argue differently) that it has to do with a few prominent people in the subdiscipline. I am not going to name names, but there are some organic chemists who are very well known and highly respected—and who are proud misogynists. And there are some physical chemists, major Nobel Prize-winning-level physical chemists, who have gone public caring about women and the status of women. I think that makes a difference; it sets a tone. Now, I don't know whether that is the whole story or if something else is going on, whether it has something to do with the nature of the subdiscipline. But it is very interesting. It would be fun to find out what other people think and what explains some of these differences.

Barbara Warren, Union Carbide Corporation: Before the meeting started, Sandra Greer and I were commenting about the fact that physical chemists have done very well compared with organic chemists. It has been discussed among women that this is because the more physical sciences are less subjective and more objective. Therefore, if you are good and you are in physical chemistry, your work speaks for itself, but some other chemistry disciplines are a little bit softer with respect to subjectivity and objectivity.

Peter Rabideau, Iowa State University: Before moving to Iowa State, I was dean of science at Louisiana State University. I put an incredible amount of pressure on the chemistry department to look for female candidates because we had no females. I think sometimes females don't want to join departments where there are no females. For example, the chemistry department responded and invited four candidates, three of whom were females, terrific candidates. We liked them all. We made great offers to all of them, and they all turned us down. So I think there is another problem—that it is just difficult when you don't have that environment. A lot of departments find themselves in the situation and are trying to counter it.

Linda B. McGown, Duke University: My impression—and this is based just on the searches that I have seen, so it is a very microcosmic sort of impression—is that in certain areas like analytical (which I think has been very open) and physical chemistry, the procedure at the assistant professor level is to go to the applicant's papers, to go to their science and evaluate it. I have noticed that among the more synthetic areas, or just other areas that tend to be more hierarchical, the way they start every search is by contacting their friends, or writing to members of the National Academy if it is for a more senior position. So the input that they are getting is from different places, and that means that they are going to get different suggestions. Now, that is very microcosmic based on experiences at one or two departments, but that first step is certainly going to structure the search.

Something interesting that I found among the same more synthetic areas is that once a woman has proven herself—once she is at the higher echelons—she becomes part of a recruitment pool. So women can become a target of opportunity. There seems to be more of a reluctance to take a chance on a female rather than a male candidate at the entry level. I have a feeling it has to do more with the cultures of the different disciplines than with the science.

Maryka Bhattacharya, Argonne National Laboratory: There is an area that I hope we can address at this meeting also. That area has to do with the big changes that we have all seen in the percentage of women who are in the student population versus their entry and retention in permanent positions in the professions associated with either university or industry. There is a lag there, I believe, where up to 30 or 40 percent of students may be female now, but when you look to where they are going and the occupations that they are going to, those percentages are changing more slowly. We need to look at what the barriers might be to retaining women in the workforce once they have the degrees, because although the student population has changed, the professions have not changed correspondingly. Why is that?

Marjam Behar, National Institutes of Health: I joined the faculty of the University of Pennsylvania School of Medicine, Department of Anesthesiology, in December 1962 to work with a group of physicians who were doing studies of cerebral blood flow, and they needed a chemist to do their metabolic studies. I didn't have a tenure-track position. As a matter of fact I was not in the faculty track, but as we advanced in the studies (I was there for 17 years), they made me director of the Core Facility for Analytical Chemistry. I had 12 technicians that I supervised and taught. I also taught residents, faculty members, and medical students who needed to learn bioanalytical techniques to pursue their research.

I was the chemist in the department, we had a center grant, and I was part of the investigators in the center grant as the director of the Core Facility for Analytical Chemistry. But I was not in the tenure track, and it is difficult for a woman in one of these universities to get in the tenure track. If you are a chemist in a department of a medical school it is even more difficult.

These are the things we go through. But even so I had a very pleasant experience doing my research, teaching, and mentoring—mainly the residents but also the technicians, who had their B.S. degrees but had the capability of going further. I encouraged them to go on to graduate school and get their doctorates in the fields of their interest. So you can make the best of a situation, but it was not ideal in the sense that you were not treated the same way as your male colleagues.

Margaret Rossiter: In fact, women sometimes find good opportunities. There are bureaucracies that have high prestige positions and people with notions of what is inferior; and then there is the real work to be done—and maybe a lot of it and maybe not well paid—and women can sometimes get into these areas and make a big contribution. But they won't have the big title, and you wonder why the bureaucracy is set up this way. If the real work is important, shouldn't that be where the focus is? And the good titles and the security?

Marjam Behar: Exactly. I continued doing my work, and I still have residents that know me and appreciate what I did and the chemistry I taught them. You can make the best of a situation, but of course it would be much better if you could be on the same level as the men.

I came to NIH in 1980, and I have found in my government career that it is better for women. We are usually at the same level as men. I would say maybe for an important promotion there is some preference for men (although we have many women who have gotten there), but in general we have the same grade as men and I think the situation is much better in government.

Suzanne E. Franks, Kansas State University: I wanted to make a few comments about trying to recruit women on the faculty and, also, keeping female students in the pipeline so that they can be recruited as faculty. One of the things I have noted over the past year in the position I have now is that search committees are often constituted entirely of men, and they don't put any women on the schedule to meet with a female candidate. So the woman is there all day being interviewed by men and never meets any other women. One might say, "Well, if you don't have any women in your department, then you cannot have women for the candidate to meet with." However, you can look around elsewhere on campus and find other women for the candidate to meet. If you are starting from this position where you don't have women on the faculty, then it is hard to recruit women—because they want to go where they are already represented. Therefore, you have to be much more creative in your recruiting strategies, and you have to make the case of selling your department as a place where a woman can come and succeed. I think women are now getting to the point where they are saying, "It is not enough for me just to get a job on the faculty. I want a job at a place where I know that I am going to be welcome, where I am going to succeed, and where there are things in place that facilitate women's success." You need to look around and find out what your university is doing to promote the success of women in the sciences and engineering and make sure you talk about them to your candidates. You need to say, "If you come here, these are all the kinds of things we have going on to make sure that this is a friendly environment for you."

I see the same thing with the women students that I talk with, and we have a very hard time convincing any of our students to go on to graduate school, especially in chemical engineering. We have a hard time convincing them to stay with their major. Some of the things they say to me are, Why do I want to be in this world? It is an unpleasant world. Why do I want to go through all these battles when I could go do something else and make more money?

I think we have to recognize that being on the faculty and reaching positions of high prestige, which we might view as very desirable, are not a priori interesting and attractive to the young women whom we might want to groom and bring into the pipeline. We have to find more creative ways of selling them on that as a viable career choice. I speak from my own experience of having gotten a Ph.D. in engineering but wanting to become a faculty member. At one point I decided to say good-bye to all that and go to industry, because I found I was treated better, shown more respect, and given more credit for the work I did. But now I am in the curious position of trying to convince women to follow the path that I chose to leave for a while—because I am not in industry anymore. I think we have to be aware that they may look at what exists out there, at what the environment is like, and say, This is not an appealing place to be. So our job in part has to be to figure out what it is that makes it an appealing place to be—and make sure that we get that message to our young women. We have to market ourselves better.

Elise G. Megehee, St. John's University: I want to speak to the last comment because last year at this time I was going through the process of looking for a new academic position. Something that really impressed me was that they didn't have any women in the department where I interviewed but they made sure that I met a woman in another department so that I could talk about women's issues. These were issues such as, What type of day care is provided? How did they treat you? and so on. The ability to ask these kinds of questions really made a big difference in my assessment of the department.

The other experience I had was with St. John's when I interviewed there. A young woman had recently had a child, and she was the only woman in the department. There was no maternity leave policy at the school, but members of the department rearranged their own schedules and worked extra so that she would have an entire semester off after her son was born. These are creative ways to get around the system, but I think it is this kind of creativity that can help change the whole attitude of a department.

Michael W. Babich, Florida Tech: We had a faculty search this year in organic chemistry, and our search committee was composed entirely of women. I thought we had the whole problem solved. It really looked like things were going to go pretty well until they returned a slate of all men candidates. So, I sent them back to the drawing board and said, Why is this? Where are the women? They showed me the applications—they had 97 applicants for the position, only four of whom were women. As far as I could tell, it was just that the odds were against the women being the best candidates—strictly a matter of numbers.

Margaret Rossiter: But there are hundreds of women getting degrees in this area.

Michael Babich: That might reinforce the point somebody made earlier. My point is, we keep seeing these numbers—that 31 percent or 35 percent or 40 percent of the graduate students are women—but they are not applying for academic positions.

Janet Osteryoung: Could I jump in here and provide some information? These data come from an NSF publication, 99-323, that reports postgraduation plans of science and engineering doctorate recipients in 1997. Those are the most recent data available. Let me preface this by saying that when I ask what percentage of women are in the applicant pool when a job is advertised in a chemistry department, the number I hear informally is 5 percent. That is the most common number that I get back, so, your figure is right on target. However, consider the approximately 70 percent of Ph.D. graduates in the field of chemistry who have definite plans—that is, their next job was already set at the time they filled out the form. For example, 30.3 percent of the men and 29.6 percent of the women have lined up industrial employment; 7.9 percent of the men and 13.9 percent of the women have academic employment; and 50.9 percent of the men and 47 percent of the women have postdoctoral appointments.

These numbers are not strikingly different. Perhaps the balance between the postdoc and academic employment shows that more women are going directly to academic employment at a non-Ph.D.-granting institution, but if you compare those numbers with the numbers for science and engineering as a whole, only one thing jumps out as being strikingly different according to gender: of this group that has definite plans, 30 percent of both the men and women in chemistry plan industrial employment, but for all science and engineering 25 percent of the men are planning industrial employment but only 14 percent of the women. It would seem, using that as some kind of normalization, that a significantly larger fraction of women in chemistry choose industrial employment.

Marion C. Thurnauer, Argonne National Laboratory: Now that we are on statistics, you showed that for graduate women the numbers have really stayed constant. In fact, there was a dip in the 1970s. I wonder if we can learn something from that, so we can at least keep the numbers constant, if not increase them. Do you know why there was a dip in the total number of chemists?

Margaret Rossiter: It might have been the environmental movement. People got turned off by chemistry. Also, I'm not sure where the economy was.

Marion Thurnauer: That is what I question. When we discuss ways to increase the number of women in the chemical workforce, how do we factor in the issues of the economy—the current job market and how it affects the chemical job market?

W. Sue Shafer, University of California, San Francisco: I want to return to the philosophy of how departments view their young, untenured faculty. When I was applying for jobs in academia (I didn't actually go there but ended up going to government), there was a philosophy that the department hired you, but you were on trial. You had to make your own way, and either sink or swim.

I recently heard Keith Yamamoto at UCSF express a different philosophy. Once you go through a search and hire a beginning faculty member into your department, it is the department's job to make sure its young faculty succeed. Instead of putting faculty on trial—ready to boot you out if you don't come up to snuff—the department views you as an investment. They need to optimize their return. A committee—like your thesis committee—would help you through the stages of getting your thoughts together, getting your first research grant, getting promoted, and getting tenured. Departments need to examine their philosophies. How are they dealing with the people that they bring into the department, and what do they want to happen?

If they want to make it a competition after the search, that is one philosophy. However, a philosophy that it seems to me would benefit most departments would be to view someone you have hired as someone you have invested in, and make your strategy for succeeding their strategy for succeeding.

Margaret Rossiter: Sometimes the department that hires isn't unanimous. So there are some who don't want you to succeed, and they may be quiet and lying in wait and then at tenure time the knives will come out.

Sue Shafer: I would just say, What a waste of everybody's time and resources.

Margaret Rossiter: They need a peace meeting after you are hired—you know, "Let's bury the hatchet; we have the future to think about. Let us not allow past resentments to color future decisions." But sometimes they harbor these resentments for decades.

Mary L. Mandich, Bell Laboratories: I would like to amplify what you said. When I was in a job search, I interviewed at a number of academic departments. Only one department offered any support for my starting career, and they told me that they would help me write my first grant. When I interviewed at Bell Labs, the people who were going to hire me said that part of their yearly job rating was based on how well I would do. That made a huge difference, and it still does.

Margaret Rossiter: It gets their attention, yes.

Christine S. Grant, North Carolina State University: I said that I wasn't going to say anything, but if you look around the room you will find that I am the only African-American person here, and so I wanted to put another statistic in your basket.

I think at present there are probably about 2200 chemical engineering faculty in the country, and I am not sure how many women. Maria Burka may be able to help me with this. Do you know about how many women there are right now—maybe about 200 or 300?

Maria Burka, National Science Foundation: About 225.

Christine Grant: Two hundred and twenty-five women. There are 26 African-American chemical engineering faculty in the country, and five are women. So, I joke with people: I say, "Me and my four friends." I have been doing this for 10 years. The other four women include an associate professor at Northeastern and an assistant professor at MIT and there is a woman at the University of Iowa and one at the University of Maryland. None of us is a full professor. These are statistics that I keep on the African-Americans and women in chemical engineering.

I met a woman at the AIChE meeting recently who said that no one had ever talked to her about being a faculty member, and she was getting ready to finish her Ph.D. So we pulled her aside and said, "You probably need to look at doing a postdoc." We find this across the board, but I just thought I would throw out that statistic.

Maria Spinu, DuPont: I was personally involved in hiring and interviewing people for CR&D, and I can testify that we looked really, really hard and tried very, very hard to find qualified women and African-Americans. We went out of our way to do that, but the statistics are just not always on our side. We did get a few women. I don't know about everything that has happened in the last 10 or 20 years, but I know what is happening now—we try very hard, and I don't think that we would pick a man over a woman who had equal qualifications.

Janet Osteryoung: I want to second what Christine said. In fact, there is a pervasive attitude about picking and choosing. It affects everybody, not just women. I think one of the things we can say is that there are so many women—and there are so many women who have received good training—that, if we cannot succeed in this, how can we succeed in anything?

2

The Advancement of Women in Science and Engineering

Virginia Valian
Hunter College and the City University of New York Graduate Center

Women are conspicuous by their absence at the most prominent levels of science, mathematics, engineering, and technology. Women scientists are sparsely represented on the editorial boards of leading journals, on the steering committees of professional organizations, and in groups like the National Academy of Sciences, as data presented in this session by Professor Margaret Rossiter demonstrated. Women are thinly represented among full professors at major research universities. Even young women (under age 35) lag behind their male peers in institutional rank and tenure.

Women in the professions are more highly represented at lower-ranked than higher-ranked institutions, spend more time in rank than men do, and make less money. In addition, women at prominent research universities have lower ranks than do women at lower-ranked institutions (with the exception of biology[1]). That such phenomena are widespread is documented in my book, *Why So Slow? The Advancement of Women,*[2] which reviews men's and women's status in the professions and academia.

Recent data from the National Science Foundation (NSF), taken together with data from other studies and other disciplines, show (1) that there is a problem, (2) that the problem is now primarily found not at entry-level positions but at later points in people's careers, and (3) that it is general across disciplines and professions—business, medicine, law, the humanities.

In 1997 (the most recent date for which data are available), women made up 25 percent of doctoral scientists and engineers at universities and 4-year colleges. That figure includes scientists who are tenured, tenure track, not in track, or in positions for which tenure is not applicable, such as postdoctoral or other appointments. Women ranged from being 6.5 percent of faculty in engineering to 59 percent of faculty in health sciences.[3] Across all disciplines, 56 percent of women were tenured or tenure track,

[1]Sonnert, G., and Holton, G. (1996). Gender Differences in Science Careers: The Project Access Study. New Brunswick, NJ: Rutgers University Press.

[2]Valian, V. (1998). Why So Slow? The Advancement of Women. Cambridge, MA: MIT Press.

[3]National Science Foundation, Division of Science Resources Studies (1999). Characteristics of Doctoral Scientists and Engineers in the United States: 1997, NSF 00-308, Project Officer, Kelly H. Kang. Arlington, VA: NSF.

compared to 72 percent of men (calculated from data in footnote 3). There was no improvement compared to 1993, when 60 percent of women in science and engineering were in tenure or tenure-track positions, compared to 77 percent of men.[4]

If the category for which tenure is not applicable is excluded, 88 percent of men overall, compared to 77 percent of women, were tenured or tenure track. The sex disparity is similar from field to field and independent of women's representation. In psychology in 1997, women were 43 percent of the faculty in universities and 4-year colleges; 89 percent of men psychologists were tenured or tenure track, compared to 75 percent of women. In the biological and agricultural sciences, women were 28 percent of the faculty; 84 percent of men were tenured or tenure track, compared to 70 percent of women. In engineering, women were 6.5 percent of faculty; 90 percent of male engineers were on the academic ladder, compared to 83 percent of women. (All calculations are from data in footnote 3.)

In 1995 (the most recent date for which age cohort data are available), sex disparities in tenure and rank are evident even for scientists younger than 35.[5] Looking just at the categories of tenure, tenure-track, and not on track, we see that 3 percent of young women, compared to 6 percent of young men, are tenured; conversely, 36 percent of young women, compared to 20 percent of young men, are neither tenured nor tenure track. As with tenure, so with rank: 4 percent of young women, compared to 10 percent of young men, are associate professors.[6] Since early career differences are limited by standard lengths of time to stay in rank, and since men and women tend not to differ in interruptions of service early in their career, the presence of small differences within the first 5 years of a scientist's career is notable.

For scientists between 35 and 44, the situation is worse: 35 percent of women are tenured compared to 49 percent of men; conversely, 20 percent of women are not on track compared to 13 percent of men. With respect to rank, only 8 percent of the women are full professors, compared to 15 percent of the men. Women even lag behind at the associate professor level, where only 34 percent of them have achieved that rank compared to 44 percent of the men. A portion of these differences may be attributed to greater interruptions in women's service. But it should be kept in mind that these figures are for full-time faculty at universities and 4-year colleges; thus, it is likely that parental leave accounts for at most 2 years of interruption. As of 1995, there were only small sex differences in number of years between B.A. and Ph.D.,[7] so that age of doctorate achievement is unlikely to play a major role in the observed tenure and rank disparities.

For women age 45 and older, the situation is grim. Of scientists aged 45 to 54 in 1995, only 40 percent of women were full professors compared to 61 percent of men; 22 percent of older women were still assistant professors, compared to 7 percent of older men. For scientists aged 55 or older in 1995, 57 percent of the women compared to 84 percent of the men were full professors; 9 percent of the women were still assistant professors compared to 2 percent of the men.

The overall picture in academia is that women start out slightly behind men in rank and tenure and

---

[4]National Science Foundation (1996). Characteristics of Doctoral Scientists and Engineers in the United States: 1993, NSF 96-302. Arlington, VA: NSF.

[5]National Science Foundation (1999). Women, Minorities, and Persons With Disabilities in Science and Engineering: 1998, NSF 99-338. Arlington, VA, Appendix Table 5-10.

[6]National Science Foundation (1999). Women, Minorities, and Persons With Disabilities in Science and Engineering: 1998, NSF 99-338. Arlington, VA, Appendix Table 5-9.

[7]National Science Foundation (1999).Women, Minorities, and Persons With Disabilities in Science and Engineering: 1998, NSF 99-338. Arlington, VA, Table 4-46.

become increasingly disadvantaged as they age. Put another way, universities and 4-year colleges are wasting the talents of their female scientists and engineers.

In industry, women scientists and engineers have fewer subordinates than do men. Women under age 35 in 1995 supervised 7.1 employees, while their male peers supervised 8.5. Between the ages of 35 and 44, women supervised 8.9, and men, 10.8. Between the ages of 45 and 54, women supervised 7.0, and men, 16.5.

It is important for everyone concerned with gender equity to know the relevant statistics. Most people are unaware of the data. Many believe that gender equity is a problem that will take care of itself as more women enter science. A full understanding of the data will help to dispel those misconceptions.

We also need to understand the irrelevance of exceptions. Everyone can think of women who are exceptions to the overall position of women; everyone can think of some very successful women. But an exception is just that—an atypical event. The fact that there are a few successful women should not distract us from the main body of evidence, which shows that—overall—women are not as successful as men, even when they have the same credentials.

Knowing the data is not enough. We need to go behind the data to an explanation. We need to understand the social cognitive processes that disadvantage women and advantage men, even in situations where the participants sincerely espouse meritocratic and egalitarian beliefs.

ANALYSIS OF THE PROBLEM: GENDER SCHEMAS

Broadly speaking, women's abilities, accomplishments, and contributions appear to be worth less than men's even when they have the same credentials (or differences in credentials are controlled for). The explanation I offer uses two key concepts: gender schemas and the accumulation of advantage. The application of gender schemas makes it more difficult for women to accumulate advantage. Schemas are hypotheses that people use to interpret social events. Schemas are similar to stereotypes, but the term "schema" is more inclusive and more neutral. The term is preferable because schemas are a necessary conceptual framework for understanding and predicting the social world and for knowing how to behave within it. Schemas are protoscientific hypotheses about social groups. We need schemas because we cannot treat every piece of data as if it is brand new and is independent of previous data. Schemas allow us to move more efficiently. They allow us to make predictions. We cannot get rid of them. At the same time, like all hypotheses they are susceptible to error. Once entrenched, they are difficult to dislodge just via disconfirmatory data. It is hard for people to revise a cherished theory, especially if it is not conscious.

Gender schemas are hypotheses about what it means to be male or female, hypotheses that we all share, male and female alike. Gender schemas are the beliefs people hold in common—whether they want to or not—about the genders. Schemas assign different psychological traits to males and females. Males are seen as capable of independent action (agentic), task oriented, and instrumental; females are seen as nurturant, expressive, and communal.[8, 9, 10, 11] In brief: men act, women feel and express their feelings.

[8]Bakan, D. (1966). The Duality of Human Existence. Chicago, IL: Rand McNally.

[9]Martin, C. L., and Halverson, C. (1987). The roles of cognition in sex role acquisition. In D.B. Carter (ed.), Current Conceptions of Sex Roles and Sex Typing: Theory and Research (pp. 123-137). New York: Praeger.

[10]Spence, J.T., and Helmreich, R.L. (1978). Masculinity and Femininity: Their Psychological Dimensions, Correlates, and Antecedents. Austin, TX: University of Texas Press.

[11]Spence, J.T., and Sawin, L.L. (1985). Images of masculinity and femininity: A reconceptualization. In V.E. Oleary, R.K. Unger and B.S. Wallston (eds.), Sex, Gender and Social Psychology (pp. 35-66). Hillsdale, NJ: Erlbaum.

The main answer to the question why there aren't more women at the top is that our gender schemas skew our perceptions and evaluations of men and women, causing us to overrate men and underrate women. The small daily events in which men get a slight advantage add up over the long haul to put them at a large advantage relative to women. There are, of course, situations in which women are actively discriminated against and harassed. But in many professional contexts, nothing overtly seems wrong. Gender schemas explain what is wrong when the problem is invisible or appears trivial; they operate on a minute-to-minute basis throughout the workday.

Experimental data demonstrate that observers do not see people simply as people, but as males or females.[12] Once gender schemas are invoked they work to disadvantage women by directing and skewing perception, even in the case of objective characteristics like height. In one example,[13] the experimenters exploited the fact that our schemas include the information that men are on average taller than women. In this experiment, college students saw photographs of other students and estimated their height in feet and inches. The photos always contained a reference item, such as a desk or a doorway, so that height could be accurately estimated.

Unbeknownst to the participants, the experimenters had matched the photographs so that for every photograph of a male student of a given height there was a female student of the same height. The question was how accurate the judges could be when faced with a sample that violated the population tendencies.

The evaluators were affected by their knowledge that men are on average taller than women; they judged the women as shorter than they really were and the men as taller. In this experiment, as is typically the case, there were no differences in how male and female observers perceived the others. That is, both male and female observers rated males as taller than females. The data from this experiment are typical: men and women make the same judgments to the same degree.

In the case of professional competence, perceptions are similarly prone to error. People are likely to overvalue men and undervalue women. One can expect gender schemas to play a role in evaluations whenever (1) schemas make a clear differentiation between males and females, and they do for professional competence as much as for height, and (2) evidence is ambiguous and open to interpretation, as is the case with professional competence. We are all tempted to think of scientific excellence as straightforward and objective, and we have difficulty seeing how much interpretation is required of the data in front of us.

A real-life demonstration of the importance of schemas comes from a study of the Swedish Medical Research Council's awarding of postdoctoral fellowships in 1995.[14] Although women were 46 percent of the applicants, they received only 20 percent of the fellowships. An analysis of the judgments made by the senior scientists on the panels showed that women received lower "scientific competence" scores than men did. To determine what contributed to scientific competence, the investigators developed a model called "total impact points," using a combination of productivity and prestige of the journal in which the young scientists had published. This model predicted scientific competence scores well for the young male applicants. But women had to receive 100 or more impact points in order to get the same rating from the judges that a man with 40 or fewer impact points got.

---

[12]Valian, V. (1998). Why So Slow? The Advancement of Women. Cambridge, MA: MIT Press.

[13]Biernat, M., Manis, M., and Nelson, T. (1991). Stereotypes and standards of judgment. Journal of Personality and Social Psychology, 66, 5-20.

[14]Wenneras, C., and Wold, A. (1997). Nepotism and sexism in peer-review. Nature, 387, 341-343.

The judges undoubtedly did not intend to discriminate against the female applicants. Nevertheless, they saw a male's qualifications as worth more than a female's. The same data were interpreted differently depending on the sex of the applicant.

Not only do schemas affect perceptions of competence, they also make it difficult for women to reap the benefits of their achievements and be perceived as leaders.[15] College students were shown slides displaying five people seated around a table. The group was described as working together on a project. Two people sat at each side and one person sat at the head of the table. Sometimes the group members were the same sex, sometimes of different sex.

The observers were asked to identify the leader of the group. In same-sex groups, the person sitting at the head of the table was reliably identified as the leader. In mixed-sex groups, a man at the head of the table was reliably identified as the leader. But in mixed-sex groups when a woman was at the head, observers sometimes labeled her as the leader and about equally often labeled a man seated elsewhere at the table as the leader. There were no differences between male and female observers. There was no intention to discriminate. The implication of this experiment is that the symbolic position of leadership carries less weight for a woman than a man. Women are less likely to obtain the automatic deference that marks of leadership confer for men. Women are objectively hurt in situations of that sort, even though observers intend no harm.

For women aspiring to scientific leadership, then, the road will be rougher than it will be for men. It will be more difficult for women to be rated as competent (as the Swedish Medical Research Council data suggest) and more difficult for them to be perceived as leaders, even when they are potentially in a leadership position.

It should be emphasized that the claim here is about tendencies. Not every woman will experience problems. Even women who do experience problems will not experience them at every point in their career. Rather, on average, women will have more difficulty than men do.

Senior members of a field, gender schemas in place, take men more seriously than women. (As it happens, most of those senior members are male, but, according to the gender schema analysis, senior women are as likely as senior men to undervalue women relative to men.) Senior members are thus more likely to pass on important information to young men than to young women (seeing them as likely to benefit from the information) and are more likely to intervene helpfully in the careers of young men compared to young women (seeing them as likely to excel). Young men are more likely than young women to be identified as rising stars and to be groomed for success.

When women do actively adopt an assertive leadership style, they are perceived more negatively than men. A laboratory study measured people's facial reactions to people trained to act as a leader.[16] The study demonstrates that both women and men—unconsciously but visibly—react negatively to women in a situation that is aimed at finding a group solution to a problem. People respond especially negatively to women's attempts to be assertive.

The researchers duplicated in the laboratory a common everyday situation in which a small group of people, in this case four people, must arrive at a decision. In the experiment the group's task was to rank, after 10 minutes of discussion, how important nine items (such as a first aid kit and a map) would be if one crash-landed on the moon.

---

[15]Porter, N., and Geis, F.L. (1981). Women and nonverbal leadership cues: When seeing is not believing. In C. Mayo and N. Henley (eds.), Gender and Nonverbal Behavior. New York: Springer-Verlag.

[16]Butler, D., and Geis, F.L. (1990). Nonverbal affect responses to males and female leaders: Implications for leadership evaluation. Journal of Personality and Social Psychology, 58, 48-59.

In order to investigate differences in facial reactions to men and women leaders, the investigators put together groups of four people. Two members of each group—one male and one female—were naive participants, undergraduate college students, who were videotaped. The other two members—also one male and one female—were upper-level undergraduates whom the experimenters had trained to be leaders, using a friendly, cooperative, and pleasantly assertive style.

The experimenters' main interest was the subtle reactions of the naive participants to females and males who were making the same suggestions in the same way. Facial reactions, because they are less under a person's direct control than oral comments, are a good, subtle measure of how someone is reacting to another person. Both male and female leaders received a certain amount of negative facial reactions; observers may harbor some resentment toward leaders. But males, unlike females, received more positive than negative reactions. Women ended up with a net loss. The male and female naive subjects did not differ in their reactions. Both sexes saw the male leader more positively than the female leader.

Thus, when women attempt to be leaders they lose, relative to men. They lose in three steps. First, they are attended to less. Women have more difficulty than men in gaining and keeping the floor. Second, when women do speak and behave in a leaderly way, they get negative reactions from those around them, even when the content and manner of their presentations is identical to men's. Men are encouraged to be leaders by the reactions of those around them. Women, conversely, are discouraged from acting in a leaderly way by the reactions of those around them. Third, otherwise neutral observers are also affected by negative reactions and tend to go along with the group judgment.

ANALYSIS OF THE PROBLEM: ACCUMULATION OF ADVANTAGE

Many of the cases in which a woman is underrated, does not receive information, does not get public notice, or is not perceived as a leader, are of small scale. It is difficult for people to appreciate the long-term consequences of small differences in treatment. Women who react to such differences may be told by well-intentioned colleagues not to make a mountain out of a molehill. That is where the notion of accumulation of advantage comes in.[17, 18, 19, 20, 21] It tells us that, piled one on top of the other, molehills *are* mountains.

Like interest on capital, advantages accrue; like interest on debt, disadvantages accrue. Very small differences in treatment can, as they accumulate, have major consequences in salary, promotion, and professional prominence.

A computer simulation of promotion practices at a hypothetical corporation provides a convincing demonstration of the cumulative effects of small-scale bias.[22] The simulation modeled an organization with an 8-level pyramidal hierarchy, in which each level was staffed with equal numbers of men and

[17]Cole, J., and Singer, B. (1991). A theory of limited differences: Explaining the productivity puzzle in science. In H. Zuckerman, J.R. Cole, and J.T. Bruer (eds.), The Outer Circle: Women in the Scientific Community (pp. 277- 310). New York: W.W. Norton.

[18]Fox, M.F. (1981). Sex, salary, and achievement: Reward-dualism in academia. Sociology of Education, 54, 71-84.

[19]Fox, M.F. (1985). Publication, performance, and reward in science and scholarship. In J. Smart (ed.), Higher Education: Handbook of Theory and Research (pp. 255-282). New York: Agathon.

[20]Long, J.S. (1990). The origins of sex differences in science. Social Forces, 68, 1927-1315.

[21]Merton, R.K. (1968). The Matthew effect in science. Science, 159, 56-63.

[22]Martell, R.F., Lane, D.M., and Emrich, C. (1996). Male-female differences: A computer simulation. American Psychologist, 51, 157-158.

women. The simulation assumed a certain percentage of incumbents would be promoted from one level to the next. Finally, it assumed a tiny bias in favor of promoting men, a bias that accounted only for 1 percent of the variance. Such a bias would typically be considered inconsequential. After repeated iterations until there was complete turnover, the highest level in the hierarchy ended up being 65 percent male. This simulation demonstrates that operating at a minute disadvantage can have substantial long-term effects.

This is the condition of ambitious women in science. They operate at a small disadvantage in a variety of areas. Those multiple small disadvantages accumulate over time to result in underrepresentation at the top.

THE ANSWER IN A NUTSHELL AND SOME REMEDIES

The basic social cognitive answer to women's slow advancement is that women's credentials do not buy them the same positive evaluations that men's credentials buy them.

Women's lower valuation is seen on a daily basis, in meetings where their suggestions are not attended to. It is seen more importantly when people look at a woman's vita and say, "I notice she published a lot of work with her mentor. Perhaps her mentor is the real mover behind this work."

The myriad ways in which we underrate women and overrate men add up over the long haul to produce women's disadvantage relative to men's. This can happen with the best will in the world, with the best intentions, with the most sincere desire to have equality between men and women in an organization.

With a firmer understanding of the social cognitive basis of gender inequity in hand, it is possible to map out some of the on-the-job consequences of gender schemas for women and men and to work out solutions. A few examples of how to approach the problem are given here.

- *Consequence and solution: Women's inadequate access to information and public notice.* As a result of gender schemas that portray women as less professionally competent and ambitious than men, women end up with less information and less public notice than do men. Compared to men, women have less access to informal routes of information and are less likely to be given opportunities by their superiors to receive public notice. For example, a recent study of the Johns Hopkins University Department of Medicine within the School of Medicine demonstrated that senior faculty were six times as likely to suggest names of junior male faculty rather than female faculty to chair conference sessions. Qualified junior women were also less likely to be identified as candidates for promotion compared with qualified junior men.[23] Thus, women are not as likely to be identified as star material, and as their careers continue they become less and less likely to be perceived as important scientific contributors to their field.

Women can press for and leaders can provide information about the criteria for success within their organization and within their discipline. Women can press for and leaders can provide opportunities for women to shine.

- *Consequence and solution: Cognitive unavailability of women's names.* When academics are choosing contributors to invite to colloquia and conferences or nominate candidates for awards, they do

[23]Fried, L.P., et al. (1996). Career development for women in academic medicine: Multiple interventions in a department of medicine. Journal of the American Medical Association, 276, 898-905.

not use a systematic search procedure. Nor, typically, are they concerned about overlooking members of any particular social group. They choose names that are cognitively available and accessible. Many cognitive and social factors contribute to determining whose names are cognitively available—recentness of mention, frequency of mention, prestige of setting in which mention occurred, recentness of last acquaintance, frequency of acquaintance, degree of acquaintance, prestige of setting in which acquaintance occurred, impression based on acquaintance, prestige of institutional affiliation, and status within the institution.

Women are disadvantaged compared to men in many of those factors. Since women do not publish as much as men, for example, their frequency and recentness of mention is lower than men's. For another example, women are overrepresented at less prominent institutions and tend to have lower ranks at more prominent ones. That, too, will contribute to their lower cognitive availability in others' minds. The factors that contribute to males' higher visibility in science also contribute to their greater cognitive availability. The effects of women's invisibility are evident in experimental psychology, where women are numerous. The June 2000 national annual meeting of a major psychology organization, for example, had 23 invited speakers, 20 male. Of a special set of 5 interdisciplinary symposia, 4 consisted only of male chairs and presenters. Here, too, no overt discrimination was intended. The organization has been in existence for 12 years, has had 6 female presidents, and had a female president when the program was developed.

Women and men can work actively to nominate women for important positions and awards. Institutional leaders can address the problem of cognitive unavailability by ensuring women's representation as invitees and awardees at rates proportional to their representation in the discipline.

- *Consequence and solution: Need to legitimate female leaders.* Leaders legitimate other leaders. In a study of person evaluation, undergraduate evaluators watched a videotape in which five graduate students had a group discussion.[24] On the tape, a male faculty member introduced one of the students as the leader. In one version of the tape the faculty member vouched for the student's expertise, mentioning the student's theoretical knowledge and performance ability; in another version the faculty member simply said the student would be the leader. The two videotapes were otherwise identical. After watching the video, the evaluators judged the student leader on a number of dimensions, including how much leadership the leader showed, how good the leader's contributions were, how desirable it would be to hire the leader, and how much salary the leader deserved.

In the tape where the faculty member had vouched for the student's expertise, the leader scored higher on all measures. The same effect occurred whether the student leader was male or female, and there was no difference in how positively male and female leaders were rated. There was also no difference in how male and female evaluators responded. A credible authority figure can successfully legitimize females as well as males. What appears to happen in such a situation is that evaluators interpret the direct information they get about a potential leader in the light of any earlier information that they have. If the prior information legitimizes the leader, the leader's behavior is seen as an example of being a good leader. If the prior information is not legitimizing, judges do not see a leader in as positive a light.

A subsequent study showed that both male and female authority figures can legitimize other lead-

[24]Brown, V., and Geis, F.L. (1990). Nonverbal affect responses to male and female leaders: Implications for leadership evaluation. Journal of Personality and Social Psychology, 58, 48-59.

ers.[25] Using the same videotaped group discussion scenes, the experimenters had one version that included a prior endorsement by a male authority and one that included a prior endorsement by a female authority. Both were equally effective in establishing students as leaders.

Thus, both women and men in authority can help competent aspiring women by legitimizing them as leaders. Leaders can use their influence to vouch for women's and men's value as leaders equally.

- *Consequence and solution: Need to identify problems.* Because of our belief that we are operating within a meritocracy, we do not scrutinize our procedures for their possible disproportionate impact on males and females. We do not ensure, for example, that men and women are aware of institutional resources and have the same access to them. But a study of men and women in academic medicine reported that, even at the beginning of their careers, men had more resources than women.[26]

Women can press for and institutional leaders can provide the tools necessary to determine subtle inequalities within their organization that help men more than they help women.

In short, we can understand the causes of women's slow advancement and do something about them.

DISCUSSION

Pushpal Murthy, Michigan Technological University: That was a scary talk. It was an interesting talk as well. I was looking for a solution, what we can do, and I heard you say that we have to depend on leaders who are going to pick us out. I agree with that, but it means we are depending on somebody else to help us. I wondered if you could address any other things where we can define the function for ourselves?

Virginia Valian: Good point. I focused on leaders because a number of you are leaders in the field and can do more than you might realize. But certainly there are things that everybody can do. Women need to work together, as the people in this room are doing, much more actively on their own behalf and the behalf of women more generally.

Ambitious women often do not want to affiliate with other women, and there are many reasons why that should be the case, not least of which is that ambitious women know that women as a group are losers in the status hierarchy. Why would one want other people to see one as part of this group?

We need to get beyond that and to understand why we might feel that way and to actively work on behalf of women in both large and small ways.

Cecily C. Selby, Radcliffe Institute for Advanced Study, Harvard University: A comment and a question. The comment relates to MIT in the 1970s, when Jerry Wiesner and Howard Johnson as president and chairman of MIT did exactly what you described in terms of finding women on the faculty. They actively promoted them and moved them, and that is how Sheila Widnall got to be Secretary of the Air Force and Shirley Jackson head of the Atomic Energy Commission. Margaret MacVicar was advanced before her premature death, and of course we all know of the continuing advance of Millie Dresselhaus. Jerry and Howard are the best example I know of being really proactive.

[25]Geis, F.L., Brown, V., and Wolfe, C. (1990). Legitimizing the leader: Endorsement by male versus female authority figures. Journal of Applied Social Psychology, 20, 943-970.

[26]Tesch, B.J., Wood, H.M., Helwig, A.L., and Nattinger, A.B. (1995). Promotion of women physicians in academic medicine: Glass ceiling or sticky floor? Journal of the American Medical Association, 273, 1022-1025.

Here's my question: Can you imagine what would to me be nirvana, a chemistry that would recognize and reflect both gender schemas? Could or would you hypothesize that the quality of chemistry would be improved by the participation of both gender schemas—and maybe some other schemas, too?

Virginia Valian: I would argue in favor of gender diversity. There is some evidence, not as strong as one would like, that innovations are more likely among a diverse group of people than among a homogeneous group of people. A more diverse workforce, diverse with respect to sex, race, class, and age (we are wasting the talents of older people) would increase the likelihood of innovative solutions to problems. The MIT example that you raised is an interesting one, because however successful MIT was at one, limited earlier time, it did not continue. We see that effort must be constant.

Cecily Selby: We learned that from Nancy Hopkins.

Virginia Valian: Right. Gender equity requires constant ongoing effort. There is no magic bullet. There is no one-time fix. We have to keep at it all the time. Unless we do, our schemas will reassert themselves and recreate the problem we thought we had solved.

Victoria Friedensen, National Academy of Engineering: I am the director of a project by the National Academy of Engineering, Diversity in the Engineering Workforce. We have done a couple of things to promote, and otherwise encourage, an expanded definition of who is an engineer in this country.

I also have an observation and a question. In a previous life I worked for NASA and did a lot of peer review management. One of the things we discovered, after a complaint was lodged, was that very few women PIs [principal investigators] were getting funded by NASA, and they were certainly submitting proposals to NASA. We took a look at the claim, and what we did was a language analysis of the committees at work. We found that in peer review committees of men and women—and this goes to your schemas—during proposal discussion, men were referred to by their titles and names: Dr. Johnson, Dr. Smith, Dr. Jones, and so on "from this institution" or "that institution." The committee member speaking of the proposal would review it and would generally use the PI's title and last name even if the reviewer knew him personally. Women would be referred to initially as Dr. Johnson and Dr. Smith; however, they were also referred to in the third person. For example, "she did" and "she is in his lab" and "she worked for him." There was a difference in language. When the reviewer brought the language assessment forward to the committee, the committee would very often go back again and reexamine the proposal. They sometimes, but not always successfully, revised their initial estimation based on their unexamined assumption and how they presented that individual. We found that the presenter might have every good intention in the world and yet was doing a disservice to the PI by using what I would call "petite" language.

My question is this: Have there been any studies done in which researchers are tracking the activities of women who are in leadership positions? Are they making a difference, whether consciously or unconsciously, to the promotion of women as they come up through the organization? Is the presence of a woman manager, in other words, going to make a difference to the kinds of schemas in promotion activities that you mentioned?

Virginia Valian: The schema data suggest that having a few women in positions of leadership isn't good enough. A woman who doesn't affiliate with other women and who doesn't know how evaluations are skewed by gender schemas is likely to make the same kinds of judgments that a man does.

That said, at workshops that I have given, I have found differences between men's and women's initial responses to a questionnaire that I give. Women are more aware of gender equity problems than men are, even if they themselves are not actively working to confront those problems and even if they do not see themselves as experiencing any gender-based problems. They are more aware than men are that problems exist, and they seem more receptive to the kinds of information that I present.

My hypothesis about what happens is that women have been tracking examples of gender inequity all along—while studiously not paying attention to them. When they acquire a framework within which to make sense of gender imbalance, all those examples get pulled in. They are then more convinced than men because men haven't been tracking those examples right along. Men don't have the same database that women do.

Women as a whole are thus likely to be a better source for changes in gender equity, as indeed the composition of this room would suggest. But we shouldn't think women are the solution and we shouldn't think men are the problem.

L. Shannon Davis, Solutia, Inc.: I have a question that builds on the previous one. We have talked much in rooms like this at other professional meetings about the impact of critical mass. At what point do you think critical mass will start to begin to impact your schemas?

Virginia Valian: That is a really good question. There actually are data on that. I will tell you an experiment that demonstrates the importance of having women in a candidate pool. Madeline Heilman conducted the experiment. She gave people a resume to evaluate along with 7 others. She varied the sex composition of the remaining seven.

The (fictitious) person you have to evaluate is always female. The other seven are either all males, six males, five males, four males, or zero males. (This is a between-subjects design.) The resumes stay the same. All Heilman varies is what the composition of the pool looks like with respect to the representation of women. Up to about 38 percent of women, women are evaluated more negatively than they are if there are 38 percent or more in the pool.

Women in business talk about the rule of three: you need to have three women in any group in order for it not to be too biased in the male direction with respect to looking at female candidates, for it to be receptive to issues that are of traditional concern to women, and for a women in a group to get a hearing for her ideas.

So the composition of the pool matters. If the pool is almost all male, it looks like a job for men. It also makes the woman look more feminine. Other work by Heilman has shown that the more feminine a woman appears, the less competent she appears to both males and females. If she is the only woman in the group, her femininity and her status as a woman are highlighted. She is seen as less competent than she would be seen if there were some other women around.

However, in those National Academy of Sciences data that Professor Rossiter showed I was particularly struck by the social science category, which had the tiniest sprinkling of women, despite the very large number of women who are getting, and have been getting, advanced degrees in the social sciences. These women are grossly underrepresented in the National Academy. Representation is not enough, but it helps.

Marylee Southard, University of Kansas: That is a good point for my question. The most disturbing statistic that you showed on your handout was on page 3 in the middle of the page. This is academic tenure across all disciplines, at universities and 4-year colleges. It compares data that are 20 years apart. This is not just in the sciences and engineering. Is that correct?

Virginia Valian: That is right. That is, everybody mixed in together.

Marylee Southard: This suggests to me that this is not merely a chemical sciences problem and that we may actually be doing better than the system as a whole. There may be some things that have been borne out in other studies about academe showing that it is not friendly and not working for women. Could you comment on this?

Virginia Valian: It is an across-the-board phenomenon. I don't know data on advancement of women in chemistry, but the other data on that numbers sheet I handed out—for women in science and engineering—certainly do not show women doing well.

I should also mention that the same database is not being used from comparison to comparison, even within the NSF data. You cannot look at the exact numbers across these different comparisons because they don't have the same denominators. (All of the NSF data are available online, by the way, at the NSF Web site. You can just print out the particular tables that interest you.)

The tenure data and the rank data are not exactly the same. The numbers differ by several hundred, sometimes several thousand. It is really hard to make clear numerical comparisons. What we should focus on is the fact that wherever you look, women are not advancing. You can look anywhere: women are not advancing at the rates they should, and it cannot be explained by things like increasing numbers of women at lower positions, because you can subtract out those older men and women who are full professors and recalculate and you still see a problem.

Women spend too long in rank, period.

W. Sue Shafer, University of California San Francisco: I recently read a fascinating book written by Susan B. Evans and Joan P. Avis, faculty at the University of San Francisco. Their book, *The Women Who Broke All the Rules: How the Choices of a Generation Changed Our Lives*, reports the interviews of 100 women in the baby boomer generation born between 1945 and 1955. Coming of age in a time of enormous social change affected both their initial career aspirations and their eventual careers. The book has a number of fascinating lessons in it, including the factors that made the difference for those women in terms of the careers they were able to achieve. One of the most significant lessons for me came toward the end of the book.

The authors asked each of these women if they were feminists, and many of them refused to accept that label. However, if you asked them if they believed that women and men ought to succeed equally they all, of course, agreed. Having avoided the label of feminist myself, it was an insight to me to understand that we have allowed that term to be taken away from us. It now carries negative connotations that we don't necessarily want to espouse. A basic concept of feminism is that all people succeed according to their ability. I think most of us would agree with that as a concept that we could support.

Janet G. Osteryoung: In the interests of keeping on time I will take the questions from the people who were standing, but please try to keep it brief.

Marion C. Thurnauer, Argonne National Laboratory: I have given a lot of thought to this issue of schema (without knowing this term). I have used different terms—socialization for one. We do get beyond the schema at certain points in time, in specific situations. I have often wondered, however, As our society becomes more diverse and we have people coming from other cultures, can we move beyond the schema? How different are the gender schemas in different cultures and how does diversity ultimately affect the outcome?

Virginia Valian: I will address your last point first. Schemas are unlikely to differ enough from one culture to another to give us any hope that we will find a culture where there is no problem with gender equity. Certainly there are cultural and subcultural differences in exactly how gender schemas are constituted and how they play out. There are odd separations in terms of things like political representation, for example, in which this country—at least at the national level—has a very poor record compared with some other countries (in part because of different voting systems). But in terms of other measures we do somewhat better than other countries.

It is also not the case that every indicator tells exactly the same story, but there is enough commonality even globally with respect to how women and men see other women and men that we can take that as a given. The increasing diversity of the workplace could work in women's favor if they were positioned in such a way that they could take advantage of it. We could try to capitalize on that.

Geraldine V. Cox, Eurotech, Ltd.: I find it interesting that we really haven't touched on the issues of sexuality. Many would-be mentors fear that they will be unjustly accused of an improper relationship when they are working with a student and trying to help a career.

I have to admit that I was pretty oblivious to this when I was in graduate school. I was very goal directed—I wanted to do my work, get out, and get on with my career. I ran into my research adviser about 20 years later and he asked, "Do you know how many faculty asked me about my relationship with you?" I was shocked because it had never even crossed my mind that his mentoring would be questioned. I think there is a real concern here, especially with male-female relationships, that many male faculty are reluctant to mentor a woman for fear of being accused inappropriately of some sort of relationship that doesn't exist.

Virginia Valian: Yes, sex itself is a factor. It is something that we should turn our attention to so that it doesn't prevent effective mentoring. Too often fears of being accused of sexual harassment or inappropriate sexual behavior are used as an easy way out. We need to say, Okay, that is a problem. Now, let's figure out what we can do to solve it.

Marjam Behar, National Institutes of Health: I think this really goes back to the societal outlook on men and women many, many years ago and the different toys that girls and boys were given to play with. It was unheard of that girls would play with any toys that they had to think about how to use—it was always a doll or a dollhouse—and the boys had other, more interesting and challenging toys. The same was true about careers. Women were thought of as going into home economics or something similar, while the men were thought of for careers in science, and that, again, changes with cultures because it depends on personal values and attitude.

My father was a man who was three generations ahead of his time. He had two daughters and taught us that the most important thing in our lives was education, and that we would have plenty of time to get married—if we wanted to get married. It was our choice, but the most important thing was to get a career. I taught youngsters, and when I had a very bright female student and I would tell her parents that she was very bright and had promise, they would tell me not to encourage her too much, because all they were interested in was for her to get a good husband. That attitude has to change. It's the same about calling a man a doctor: if a man is a doctor they usually don't call him mister; however, a woman doctor is usually called miss or missus. So, there is a general societal attitude also that we have to consider.

Virginia Valian: Sure, there are problems everywhere. But I don't want us to make the problem so big that we give up and say there's nothing we can do. It is important for us to specify particular areas where we see we can make a difference and make a difference in those areas. Everyone needs to work, in his or her own way, on making gender equity a professional reality.

3

Workforce for the 21st Century: The Federal Perspective

Arthur Bienenstock
Office of Science and Technology Policy

As I looked through the program, it appeared that you are appropriately focusing on individual and organizational aspects of the participation of women in the chemical sciences, although it is likely that the issues you are considering apply to all of the physical sciences and engineering. Those issues are real and occupied my attention for an extended period. I was Stanford's first faculty affirmative action officer, supervised a significant number of women graduate students, and even altered Stanford's policy on the admission of women many years ago.

Today, however, I would like to take a completely different tack. That is, I will focus on the entire national scientific, technical, and engineering (ST&E) workforce from the perspective of the Office of Science and Technology Policy (OSTP) and the National Science and Technology Council, as represented in the report *Ensuring a Strong U.S. Scientific, Technical, and Engineering Workforce in the 21st Century*. That report was prepared by the Interagency Working Group established to assess the ST&E workforce and develop recommendations on how to ensure its strength; it was co-chaired by Martha Krebs (director of the Department of Energy's Office of Science) and me.

Let me first talk about the ST&E workforce issues that are most evident in the news. First of all, for the technical fields and, actually, for almost all who are well educated, there is a very low unemployment rate—an unemployment rate below that which many economists thought feasible with normal job turnover. In addition, our newspapers frequently publish statements that the unavailability of workers is limiting economic growth in one way or another. Consistent with both, there are constant calls for increases in H-1B visa quotas.

These three phenomena speak to a shortage in the United States of scientifically and technically trained workers, leading us at OSTP to ask whether these phenomena are short- or long-term trends. We cannot, of course, tell the future, but it is our obligation to make an educated guess. In doing that, the first thing we asked is, What has happened to the ST&E workforce historically? We were fortunate to have a very good intern, James Buxbaum, in the office to address that question. In particular, we asked him to determine what happened historically to the fraction of the workforce involved broadly in science and technology, aggregating all job classifications from technicians to Ph.D. scientists. His conclusion was that the fraction of the workforce involved in science, technology, and engineering grew from roughly 11 percent in 1962 to 15 percent in 1995.

That conclusion is consistent with the observations of those of us who lived through that period. Our everyday life has changed markedly through increases in the fruits of technology that have emerged from science. For example, I find it wonderful to be able to fly around the world at a rate that was impossible when I was young. There couldn't be that many airplanes up in the sky 40 years ago because the computers weren't available, nor were the jets. Then there is music. I can remember constructing my first hi-fi system in the mid-1950s. Now you see people with CD players strapped to their waists and getting higher fidelity through the headphones than anything most of us could obtain in that period. That is the consequence of the laser, the transistor, the integrated circuit, and all sorts of sophisticated digital processes. I cannot help but observe, as well, people roller-skating on the beautiful rollerblades that have resulted from the development of plastics that were unknown when I was a child—we used metal-wheeled roller skates that were far less satisfactory.

Clearly, the world has changed technologically and the fraction of our workforce involved in science and technology has increased. The next, obvious question is, Do we expect that trend to continue? Our sense is yes. One indication comes from Bureau of Labor Statistics projections of where we will need more people over the next decade. Almost all of the fastest-growing positions are related to information technology or health care. Those projections imply that it is likely that the fraction of the workforce devoted to ST&E will increase. There is another argument that can be made.

We know from all of the discussion of Social Security that our population is aging and that the ratio of workers to retired people is decreasing. That is why we worry about the future solvency of Social Security. But we can consider the broader implications of that change and conclude that if we don't have increases in productivity per worker, the standard of living will decline for everyone. The stock market will drop in value, the dollar will drop in value relative to other currencies, or there will be inflation. Something will happen to cause the average standard of living to decline if we cannot increase productivity per worker.

How do you increase productivity per worker? It's primarily through science and technology. You often hear the President, the Vice President, and many economists say that science and technology have accounted for over one-half of the increase in productivity per worker over the last half century. So we are going to need more scientists, engineers, and technicians, and everyone has a stake in that happening—even people like me who are going to retire in a few years—if we want to see our standard of living maintained, not to speak of the benefits to health, to the environment, and to national security.

With that in mind, we at OSTP have looked at other demographic projections. Figure 3.1 shows the fraction of the population that is in the age group 18 to 64 years, which we use as a surrogate for the workforce, projected over the next 50 years by the Census Bureau. The different curves represent projections for various subgroups of the population. The top two curves, representing the fraction of the population that is non-Hispanic white male and female, are practically superimposed, as you would expect. Both show a steady decline. What the graph does not show is that the group that has traditionally provided most of our ST&E workforce non-Hispanic white males—is projected to decline in absolute as well as relative size after 2010. Increasing markedly is the Hispanic portion of the population. Increasingly somewhat less markedly is the Asian-American portion of the population and increasing somewhat is the African-American portion of the population.

Why is there cause for concern? As I said, traditionally, and particularly in the fields that interest you today, non-Hispanic white males have comprised the dominant fraction of the ST&E workforce. Their fraction of the population is expected to decrease while groups that have participated at lower rates in the ST&E workforce are projected to become a significantly greater portion of the population.

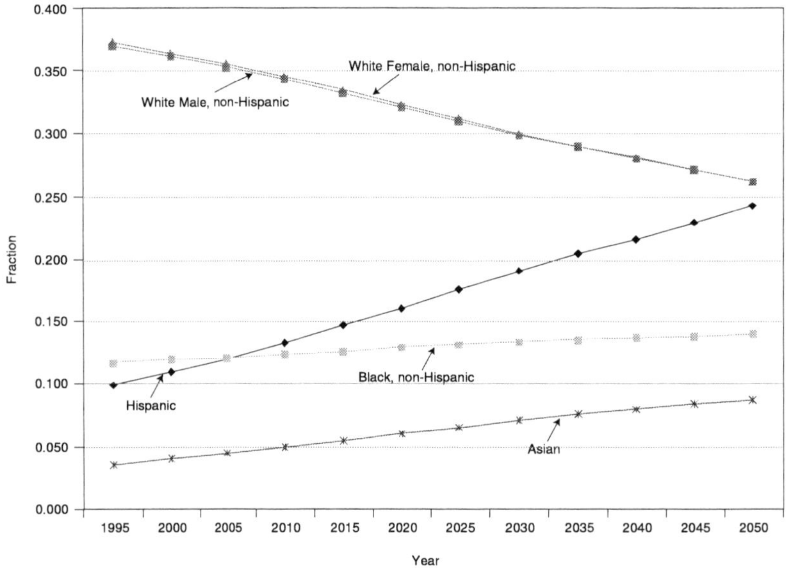

FIGURE 3.1 Bureau of the Census population projections for five ethnic and gender groups, ages 18 to 64 years. SOURCE: Day, Jennifer Cheeseman, *Population Projections of the United States by Age, Sex, Race, and Hispanic Origin: 1995 to 2050,* U.S. Bureau of the Census, Current Population Reports, pp. 25-1130, U.S. Government Printing Office, Washington, D.C.

To get some sense of the differences in participation of the various groups, consider the statistics on ST&E bachelor's degrees in 1995. In particular, consider the ratio of the number of 1995 ST&E bachelor's degrees earned by members of each population subgroup to the number of 22-years-olds in that group, expressed as percentages. That percentage is 13 for non-Hispanic white males, 11.8 for non-Hispanic white women, and 21.6 for Asians. But for the African-Americans and the Hispanics who will make up close to one-half of the workforce by the middle of this century, the numbers are 5.7 and 4.8, respectively. Those rates of ST&E bachelor's degree acquisition are significantly less than half those of the non-Hispanic whites.

For women there are two trends. First of all, there have been significant changes since I was a student, in that women now earn almost half of the bachelor's degrees in this broad area. That fraction declines when advanced degrees are considered. In addition, detailed analysis indicates that women are greater participants in the social sciences and biological sciences, while their relative participation is much smaller (about 20 percent) in mathematics, computer science, and engineering. Yet these latter fields contribute heavily to economic growth, to improving the environment, and to the maintenance of national security and our citizens' health.

Now, the interagency working group I referred to earlier asked the following question: What will happen if the nation does see the demographic changes projected by the Census Bureau and the fraction of young people earning degrees in science, technology, and engineering remains what it is now for each of the ethnic and gender groups that I have discussed here?

Our calculation indicates that there would be a steady decline in the fraction of 22-year-olds receiving bachelor's degrees in ST&E (see Figure 3.2). The calculation indicates that the increase in the fraction of Asian-American 22-year-olds is not sufficient to compensate for the decline in the fraction of white males or the low rate of acquisition of ST&E bachelor's degrees by African-Americans or

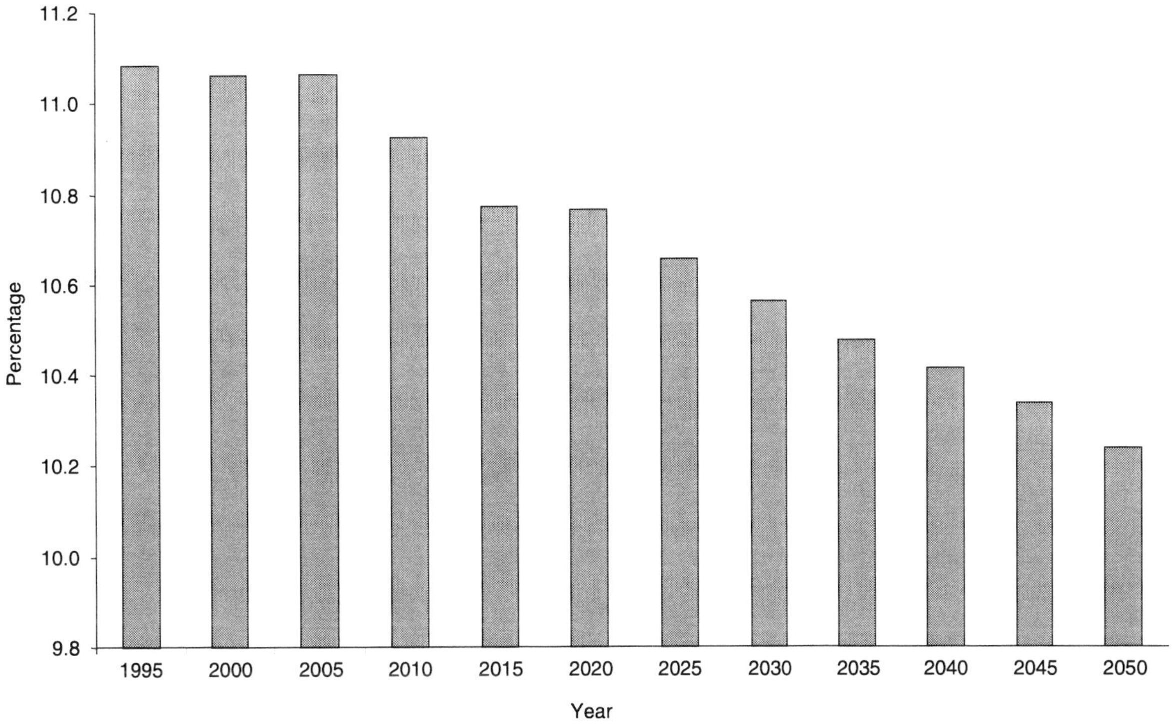

FIGURE 3.2 Calculated fraction of 22-year-olds receiving bachelor's degrees in science and engineering if award rates of various groups remain constant. SOURCE: Day, Jennifer Cheeseman, *Population Projections of the United States by Age, Sex, Race, and Hispanic Origin: 1995 to 2050,* U.S. Bureau of the Census, Current Population Reports, pp. 25-1130, U.S. Government Printing Office, Washington, D.C.; National Science Foundation, *Women, Minorities, and Persons with Disabilities in Science and Engineering: 1998,* NSF 99-338, Arlington, Va., 1999, Table 5-22.

Hispanics in maintaining the total rate of ST&E bachelor's degree awards. A decline in that rate would, in turn, be likely to lead to a decline in the fraction of the workforce in ST&E over a period when the nation will probably need a significant increase.

Moreover, it seems likely that Asian-Americans will not continue to enter the ST&E workforce at their present rate, since they are likely to branch into other professions. We can already see that happening in the Asian-American community. Its members are writing novels and plays, becoming lawyers, and competing quite successfully internationally as figure skaters. They are doing all the other things that you expect of an educated middle- and upper-middle-class community that is integrated into our society.

So we have reason to be concerned as a nation. This calculation indicates, for example, that the requests each year for increases in H-1B visa quotas are likely to continue into the indefinite future if we don't turn this situation around.

On the other hand, the countries from which we are getting immigrants clearly see the value of the people that they are losing to the United States. As a consequence, there are efforts in almost every technologically advanced Asian country to establish first-rate graduate schools and to establish positions within the country that will keep their best and brightest in the country rather than have them emigrate.

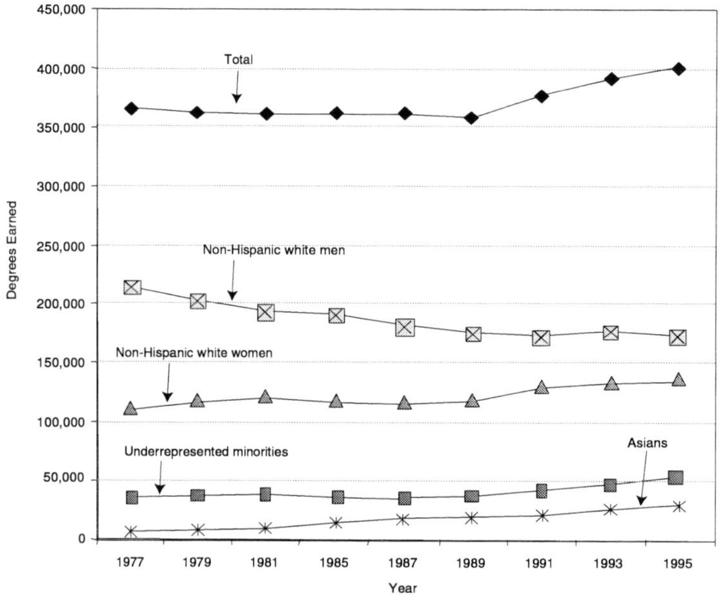

FIGURE 3.3 Earned B.S./B.A. degrees in science, technology and engineering fields, by race/ethnicity, 1977 to 1995 (U.S. citizens and permanent residents). SOURCE: for 1977 to 1991, National Science Foundation, *Science and Engineering Degrees, by Race/Ethnicity: 1977-1991,* NSF 94-306, Tables 1, 2, and 3. For 1993 to 1997, National Science Foundation, Division of Science Resources Studies, *Science and Engineering Degrees, by Race/Ethnicity of Recipients: 1989-97,* NSF 00-311 (Author, Susan T. Hill), NSF, Arlington, Va., 2000, Tables 4, 5, 6, and 31.

What conclusion do we draw from these considerations? The first and strongest conclusion is that we must increase the participation rates of all ethnic and gender groups in science, technology, and engineering. The administration's drive to improve science and mathematics education is one aspect of this effort. There are many other things that can be done. But while the general society debates the merits of affirmative action on the basis of apparently conflicting goals for justice toward individuals, we should all be asking the broader questions: What kind of country do we want to have 20 or 30 years from now? Do we want our country to be increasingly dependent on immigration to meet its workforce needs, or do we want our citizens participating more actively in some of the most interesting, productive, and rewarding ventures in the society? I think when you ask it that way, you see that you really want to encourage all Americans to participate in science, technology, and engineering. At the same time, it is important that this country remain attractive for immigration.

To ensure broader participation of our own citizens in science, technology, and engineering, while keeping the nation attractive for immigration, it is important to keep our institutions of higher education strong. Thus, the administration has been reviewing the government-university research partnership and has been working with the universities to improve a relationship that has degraded over the past decade.

Most of you probably feel, from your personal observations, that the nation has made significant progress in broadening participation in the ST&E workforce. That is illustrated vividly in Figure 3.3, which shows the number of ST&E bachelor's degrees earned by members of various ethnic groups over the past 20 years up to 1997, by year. Over that period, the number of non-Hispanic white males earning bachelor's degrees declined markedly, with the decline driven primarily by demographics. Nevertheless, the total number of degrees awarded to United States citizens or permanent residents has increased. That increase is due to three roughly equal factors: the increase in women earning bachelor's degrees, the increase in Asian-Americans earning bachelor's degrees, and the increase in African-Americans, Hispanics, and other underrepresented minorities earning bachelor's degrees. These three roughly equal effects more than counterbalance the decline in the number of non-Hispanic white male awardees. Had those increases not occurred, it is likely that the White House and Congress would have faced even greater demands for increases in the H-1B visa quotas.

Unfortunately, that progress is threatened (see Figure 3.4). Enrollments of African-Americans and Hispanics at the major universities that educate many of our scientists and engineers are declining. We have good reason to be concerned that the recent court decisions and referenda, including Proposition 209 in my home state, California, will turn back the clock. I really worry about that, and I think we all have to ask what kind of United States we want our children and grandchildren to grow up in. Will it be one in which people from all races participate in science, technology, and engineering, or will we become like some developing countries that are divided markedly into an upper class and an upper middle class that are primarily white and Asian-American, and lower classes that contain the other ethnic groups? I know my choice. I have seen what it is like in those countries, and I certainly don't want my grandchildren growing up in that type of society.

Now, let me return to the main themes of this meeting. We will not see increases in participation of women in the ST&E workforce unless we continue to address the questions that are the focus of this

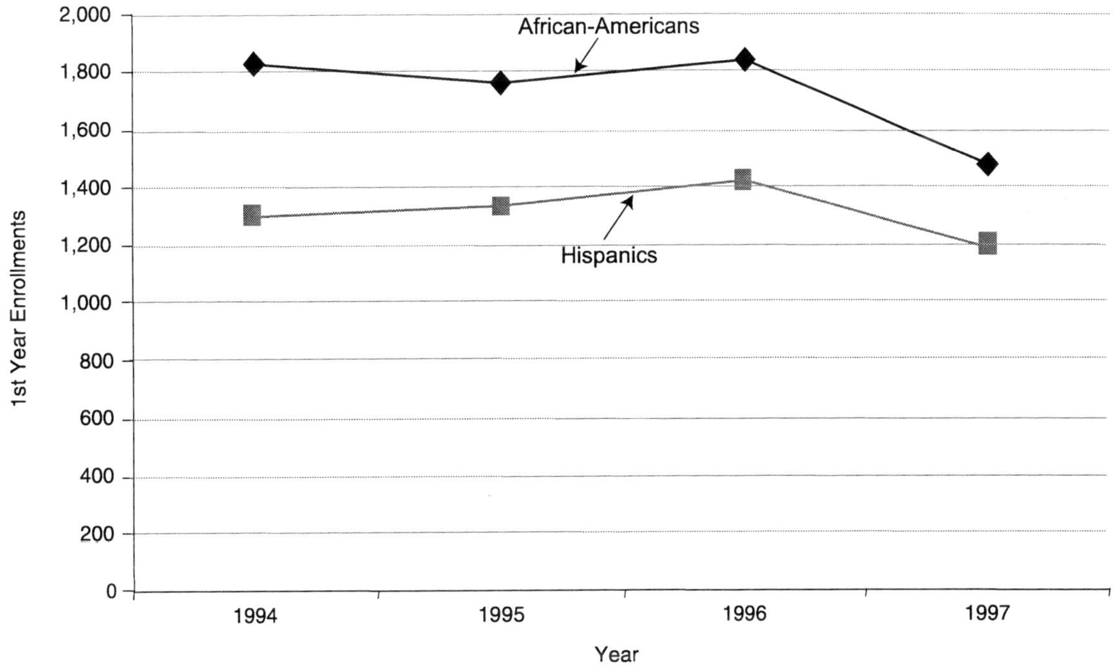

FIGURE 3.4 Entry into science and engineering graduate work declines in 1997 (first-year enrollments at surveyed universities). SOURCE: National Science Foundation/Division of Science Resource Studies, *Graduate Students and Postdoctorates in Science and Engineering: Fall 1997,* NSF 99-325, NSF, Arlington, Va., Tables 27, 28, and 29.

meeting's agenda. Part of the solution to our nation's ST&E workforce needs involves changes in our organizations, so that they can accommodate and encourage women in the professions in which they are sorely underrepresented—physics, chemistry, mathematics, and engineering. This nation needs greater participation of women in these fields. The issues that you are addressing are central to attracting them.

DISCUSSION

Geraldine L. Richmond, University of Oregon: Artie, thank you for your comments. Your last comment was one that hit a nerve; you said to this group, Unless you work on these issues you will not have a solution.

Arthur Bienenstock: That is a very good point. All of us have to work on these issues. You are absolutely right, and you are right to call me on it.

Victoria Friedensen, National Academy of Engineering: Dr. Bienenstock, I know that we are approaching an election year, and I know that your report has certainly been making a splash—at least here inside the Beltway. I know you don't like to look into the future too hard, but do you believe that the next Administration will carry on your work? Will the recommendations made in the report be carried forward and address implementing solutions to the problem rather than just restating the issues that contribute to the problem?

Arthur Bienenstock: I am a political appointee and therefore can only speak for one of the two candidates. I would anticipate in the area where I have knowledge that this issue will continue to be a big one.

Let me say that I think there was not a tendency to focus on these issues of the relationship of race and gender to the science and technology workforce until recently, and you must recognize that even in putting together the working group we were reacting to the referenda and the judicial decisions more than we were thinking what to do next, but I can assure you it has become a big issue in the White House. Two or three weeks ago, the President had an event where 25 companies each pledged the expenditure of $1 million each year for a decade—so, $10 million each—for increasing the participation of minorities and women in science and technology. Many of those companies told me they will contribute much more and indeed talked about some $6 or $7 million, and now we have other companies that want to join in and make the pledge, too.

The science, technology, and engineering workforce has become an issue within the One America Office that deals with race. Now there are close relationships between the two offices, and I have every reason to believe that the candidate with whom I am familiar would continue that.

I just cannot speak for the other one because I don't know him as well.

Kimberly Gray, Northwestern University: I have two comments. First, you put up a very troubling picture, and yet it seems to me that we don't see the same sort of concern if we look at budgetary priorities and the amount we invest in our total budget each year in science and technology and research. Even though NSF has requested a very large budget increase I don't think it is going to be successful partly because of the caps—I guess budget caps imposed by Congress.

My second concern, though, is that what we then see may be mirrored in universities. Northwestern is a university that is sometimes regarded as fiscally conservative, a university that regards science and technology as kind of expensive. It is much more economically feasible to educate someone in the humanities or even social science than it is in science and technology. We see, at least at the university level, this notion that it is just so expensive to educate scientists and engineers.

Could you comment about what we might expect with respect to budget?

Arthur Bienenstock: Again, I can speak most authoritatively about the President's budget and the President's goals in presenting the budget. He outlined those in a speech at Caltech; let me read one section of that speech (I just happen to have it with me): "This budget makes research at our nation's universities a top priority, with an increase in funding of more than $1 billion. University-based research provides the kind of fundamental insights that are the most important building blocks of any new technology or treatment. It also helps produce the next generation of scientists, engineers and entrepreneurs. We are going to give university-based research a major lift."

The budget is consistent with that. It has a 17 percent increase for the National Science Foundation, and if you look at the budgets of the individual sectors—mathematical and physical sciences, social sciences, biological sciences—you find that there are increases between 15 and 20 percent. There is an initiative within that budget to do research on the participation of women and minorities in science and technology. There is the commitment there. There is a corresponding increase of 13 percent, as I recall, for the Department of Energy's Office of Science.

What is required next is the will to use public funds to achieve those gains. The big question is, How will you disburse the funds that have been acquired through balancing the budget and that now form a surplus? Will you disburse them to be expended by the general populace, or will you use them for investment in the society? That issue divides the parties, and we will fight hard to achieve that budget.

If you believe in it you ought to be fighting hard. There are some who say that such investments are not a proper government function, but I don't think there is anyone in this room who believes that such investments are not a government function—and I have to remind you that it has been known for 3000 years that governments should use periods of plenty to save for periods when things are rough. It appears in the Old Testament in the story about Joseph. Remember, Joseph is called to interpret the pharoah's dream and urges the pharoah to set aside the grain during 7 years of plenty to deal with the tough years that are to follow. The equivalent of that in our society is investing in research and investing in people. I hope the Congress will see that and come through, but you have got to help.

W. Sue Shafer, University of California at San Francisco: In this judicial climate, how can the major universities, state or private, reverse the trends that you talk about—of losing the participation of underrepresented minorities at their institutions?

Arthur Bienenstock: Let me talk first of all about Stanford, where I was chair of the Undergraduate Admissions and Financial Aid Committee in the period after Martin Luther King's death. We really looked at what criteria you should use for admissions, and there is no way that we would use the simplistic numerical criteria that are used by so many state institutions. You miss out on really promising students if you do that. I think that the state schools, and certainly the private ones, which have a great deal of freedom, should be looking at the structure of their undergraduate classes and not be using simplistic criteria like GPAs and SATs for admission.

If you read that book by Bok and Bowen[1] you can see that the elite institutions are using more complex criteria, and it really pays off in the nature of the class that you get. I can talk about students

[1]William G. Bowen and Derek Bok. 1998. *The Shape of the River: Long-Term Consequences of Considering Race in College and University Admissions*. Princeton, NJ: Princeton University Press.

that we got at Stanford that we wouldn't have gotten otherwise. That is one thing. Second is restricting the private schools. Gerhard Caspar, as president of Stanford, said that there is no way we are turning back on affirmative action. We see it as a societal necessity. You are all citizens. Many of these things have been accomplished with referenda, but referenda can go both ways. You can state what you believe in and work for what you believe in.

We in the government are looking to private-public partnerships. It was no accident that we went out into industry, because we ourselves are limited in what we can do, but working with the private sector we can do a lot more, and we know that the high-tech industry shares our concerns about the workforce and is prepared to make efforts to change things.

So, there are many ways of doing it, but never forget that you are citizens—and citizens of your universities, too—and that you have a variety of options.

Robert L. Lichter, Camille & Henry Dreyfus Foundation: First of all, thank you for taking a leadership role in this broad issue and being so articulate about it. I hope you will continue to carry that message.

I am looking forward to seeing what the report itself says. At the same time, I am still a little bit troubled because there have been reports on this and related issues. We have seen them pile up. I get about one a week on one or another topic, as I am sure most people here do, too. Sometimes the reports almost appear as ends in themselves. The potential for this is one thing that troubles me.

The second thing is the language used to discuss this issue: the workforce argument. It certainly is true that science and technology drive and make a difference in our contemporary standard of living, but science and technology don't exist in a vacuum. It is *people* who really make that difference. What I rarely see in any of the reports that focus only on the notion of the workforce (which many private corporations say they can deal with better than universities can) is the notion of *leadership*.

What I don't see discussed are efforts to make available the massive amounts of money called for by individuals and corporations—with the attention being on the money itself—for developing leadership. I don't see the question asked, not of how *much* money should be made available but *how* that money should be used to develop leadership. I would like to hear more about this.

Arthur Bienenstock: Okay, let me talk a little bit about things that we didn't make a fuss about. First of all, let me say something that is surprising, given my elitist university background. I have become quite respectful of the role of community colleges in our educational system and also enamored of programs run by the National Science Foundation and the Department of Education that seek to provide curricula and training in technical fields where industry needs workers. They tend to be cooperative programs in which industry works with the community college system to improve curricula. There are also programs that require the curriculum to allow for transfer to a 4-year college. So a student at the end has the option of going into employment for, say, one of the semiconductor processing companies in the Southwest, or, alternatively, going on to a 4-year school.

We increased funding in the budget for those programs; we will see if they last. And we increased funding for a variety of programs of that sort. There wasn't a big fuss about it; it was just done. That was the internal leadership; then there was the calling together of larger industries, and we will continue to do that—to discuss what we can do together in government and university partnerships, at least through this Administration. But I think it is an endeavor where there have to be a lot of leaders, because there is a lot to be done when you think about it. I mean we need church groups and social groups encouraging young people to go into science and technology. It has got to spread down into the society as a whole. There are lots of different levels of leadership that are required, and Bob is right to draw attention to that.

Elizabeth C. Theil, Children's Hospital Oakland Research Institute: I would like to return to Professor Richmond's comment in the very beginning, because even if we accept this problem as our problem as women, we are a small number. We could certainly have a bigger impact if we could convince more men that it was their problem as well. What is the single most effective thing you think we could do in that regard?

Arthur Bienenstock: I always think in terms of structural reform for women. Let me say something about my experience as a thesis adviser on materials science and applied physics. Roughly half my Ph.D. students over the past two decades have been women. One of the constant lunchtime discussion topics is, How do I deal with bringing up children and having a career? We have made progress in helping families to deal with some of the issues, like child care centers, but I think we have to think as a society that desires the contributions that come in both ways—the parenthood and the professional contributions—and look for structural changes. Some universities have adjusted, changed tenure clocks by stopping a tenure clock for child care for both the father and the mother, but I must say in the end I never had satisfactory answers for my students.

I would bring in women I knew who already had successful careers, and they discussed how they do it. I think we have got to look at the issue of what structural changes could be made so that a family can be both a parenting family and a working family with both parents committed to very demanding careers. There are many careers where you work nine to five and then you leave. You can fit children into that, but it is much harder when you want to be a chemist or a physicist, because the intellectual commitment goes on all the time. I am sure you all know it, and I think we have got to think of societal changes.

You are going to be watching your own institutions. The things that you demand in terms of justice are things that will in the long run benefit all of us. It is those things, such as helping women to manage those dual roles, that count.

On the other hand, I recently discovered a shocking thing. I just turned 65, and my students threw a big party for me, which included my ex-students coming from around the world to attend. I realized that not a single woman who has gotten a Ph.D. with me has had a child, and it was a striking observation. I think that is an issue we really have to explore, and it bothers me very, very much that women's organizations that are very concerned with the role of women in science and engineering don't speak more about this issue from many different points of view. There is the obvious one—the father playing a greater role in parenting—but there have been a variety of solutions that families have achieved that have allowed women to pursue their roles in science and engineering successfully. I think our young women should hear from the women themselves and not from me.

Geraldine V. Cox, Eurotech, Ltd.: Actually this was a point I was going to make, that right now our values are changing in society. A survey that was just released said the generation now coming out of college values family more than career, and they are choosing careers that allow them more family time. That is both male and female, but I think there is also a loss of the traditional work ethic, and I have talked to a number of professors who say, "I am really disappointed with the students I am getting now. If they work 30 hours a week, they consider me being overbearing."

I remember back when I was a graduate student we used to pull all-nighters just to get the work done. That level of commitment to career has a negative value in today's society. We are not creating a sense of having to work to achieve. We are a society of instant gratification: "If I cannot have it right away, I don't really want to work for it." I think there is something at the core of our societal values that really impinges on what we are dealing with here. Career and family are both important, and I agree with you we have to work more on that, but I also think that there is a greater societal drifting away from the old work ethic that used to be part of this society that made it great.

Arthur Bienenstock: I have to say that I didn't notice that. I left Stanford to come to Washington $2^1/_2$ years ago and my students are still working like crazy.

Geraldine Cox: I am not a professor, but I have talked to a lot of professors and they were complaining about the fact that the students aren't putting in the effort.

Arthur Bienenstock: I have heard that complaint. It hasn't been consistent with my own observation.

Janet G. Osteryoung: Did Socrates complain about that?

Arthur Bienenstock: Yes.

Christine S. Grant, North Carolina State University: First of all I would like to applaud the efforts to get industry involved with minorities and women in terms of supporting them with this million-dollar initiative. I was a product of General Electric's program to increase minority engineering graduates (PIMEG) that started when I was in junior high and continued through high school. We actually got to visit engineering plants and universities. I also participated in summer programs to introduce minorities to engineering (MITE) at different universities (Northeastern and the University of Massachusetts at Amherst), spending 2 or 3 weeks on campus. I loved those programs, obviously, but let me tell you what is happening at the universities now. Maybe the following activities have trickled down from what is happening in the California [state university] system.

I award scholarships for my department, and I am on the Scholarship Committee for the College of Engineering. The university has said that we can no longer award scholarships that have anything to do with race. I don't know if this is coming for women also, but PIMEG and the MITE programs are becoming more diluted and some of them are going away or becoming shortened. I would encourage the company initiatives mentioned by the speaker to increase the participation of these groups in science and engineering.

I think it is very important for the companies to realize that they can't just give money to a college of engineering and think that it is going to support our underrepresented undergraduate students anymore. They are going to have to find more creative ways to get that funding to the students through organizations such as the Society of Women Engineers and the National Society of Black Engineers, with scholarships through those programs that can impact our students directly. Other than that, our students are never going to see the impact of that money that is being given by the industry folks.

Arthur Bienenstock: I should say that about 7 or 8 months ago in the White House, we very quietly had a meeting of the relevant people from industry—human resource officers, people dealing with these issues—where we just sat around the table and went over what is working and isn't working. We intend to do that more and more just because of this sort of situation, and we are facing it ourselves. We had to review every program that we had to see whether it was constitutional and would hold up under review. So, we are going through that, and that is what has me more worried. I mean things could just get worse, and we have got to fight back. That is why we did the report, to help fight back.

Nancy H. Hopkins, Massachusetts Institute of Technology: Thank you very much for your comment about women and children and the need to restructure the job to address the problem of the greater family responsibilities that so often fall to women. At MIT we did a study that shed light on the underrepresentation of women in science and found two major reasons that we believe may contribute to their departure from the profession. One is this subtle gender bias, and the other is the family issue. Young women faculty told us that until the latter problem is addressed there will never be a large number of women on our faculty in science and engineering. This problem is beginning to impact young men as well, although still to a much lesser degree. When I think about this problem of balancing family and work, it seems to me that the cost of fixing it may be substantial. I am wondering whether you have any suggestions on how we might go about obtaining the needed funds. How does one make people understand that we may need to put day care in every laboratory that is built, for example, and make day care an item in the budget of a grant? That we need to have housing near the campus so that people do not have to spend hours a day commuting to work and picking up their children? I think when we women left the home we forgot to note that it probably costs about $100,000 or even $150,000 a year to replace a college graduate in the home, since many of these women were working 2 or 3 shifts per day performing diverse and important jobs for the family.

Arthur Bienenstock: That is a very good point, and I don't have all the answers. That is why I would like to see more discussion of it. I can tell you what we did in the places where I have experience. Certainly there was the establishment of day care centers, but you find day care centers are pretty expensive, and what the university did was give the land and the buildings away; nevertheless, staffing is expensive and we have to accept the fact that for a period, family income is going to go into child care centers.

I mentioned delays in the tenure clock. We also introduced faculty leave programs, not for this purpose but for the purpose of giving extra time to faculty members who had been very heavily involved in things. What we found was that there were inordinate demands on our minority and women faculty. Not only were there the normal demands of teaching, research, and committees, but there were so few of them that they were being asked to provide guidance for women and women's groups and all that sort of thing. So what we did was extend the sabbatical period beyond normal for people who faced unusual demands, to give them more time for their own research quietly.

All sorts of little things like that help. They don't solve the problem, but every little thing helps a bit. Those are the things we tried, and still try to maintain. I don't have all the answers, but these are things that I think should be the subject of discussion.

4

Reports from the Breakout Sessions

Following the presentations described in Chapters 1 to 3, breakout sessions were organized to enable more extensive discussions among the workshop participants. The following questions and statements were suggested to the breakout groups as possible topics for discussion:

- Tell some stories about your workplace.
- What gender schemas do you see operating in your workplace?
- What are the consequences of underrepresentation of women in science and engineering, especially at higher professional levels?

Discussion leaders from the breakout groups then reported in plenary session what they believed to be important ideas and topics that had emerged in the discussions.

Sandra C. Greer, University of Maryland, College Park: In our group, which consisted mainly of chemical engineers, we addressed the first two of the three questions, and we started out with some stories about our lives as scientists and engineers. Some of us have been in this business for 30 years or so, and the war stories that we've told in our time are getting a little wearying. Instead, we each attempted to tell a story that had a positive outcome and see what could be learned from that in terms of both strategies and tactics to help us to move forward.

Let me talk first about the tactics or short-term lessons from these stories. One was the notion of networking in a very broad sense—networking in terms of students talking with their peers and finding inspiration and support in their pursuit as underrepresented groups.

One story described a corporate culture in a company where new scientists and engineers move around in the organization and get to meet many different people. This could be particularly valuable for women to get their faces seen and get to know people around the company.

Another way of accomplishing this kind of networking is through professional meetings, and our discussion focused on small meetings like the Gordon Conferences. These are very different from large national meetings, where you are a face in a crowd of 20,000. I remember going to my first AIChE meeting—which wasn't that long ago, because I am a convert from chemistry—and even at my age it was still intimidating to go there and be surrounded by 20,000 men.

The argument was made that small topical meetings facilitate better networking. At Gordon Conferences, you sit down at a table with people, talk to them one meal after another for a whole week, and they actually get to know you and you get to know them. Then, miraculously, when these people see your proposals and papers, they look better than they would have had you not had the opportunity to meet them. This kind of networking is a very valuable aid—a sort of short-term tactic—at all levels, for both students and senior professionals.

Another topic that we discussed extensively is inclusive language. We talked about how many of us have fought this battle repeatedly: the battle of "he or she." It really matters to keep saying "he or she," even though many people have thrown up their hands in defeat. Among chemists and chemical engineers, this has been a hard fight. Many of our colleagues won't use this language—at least many of mine just won't do it. I confess that recently I have not put as much energy into this as I should, but I will return to my institution with renewed commitment to inclusive language.

Other stories in our group led to longer-term strategies. One of those is having leaders who care. In my own department, when Jan Sengers was chair of chemical engineering at Maryland, he made it one of his high priorities to diversify the department. A long-time practice in academia is that you never hire one of your own students—your student finishes and is expected to find a job somewhere else, but certainly not in your own institution.

We had a fine student, a young African-American woman, who was a chemical engineer. There are five of those in the country, as we learned earlier. Jan worked out a plan that allowed her to go elsewhere for a postdoc, but then we offered her a job. She came back to Maryland, where she is now a faculty member and has just won an NSF CAREER award.

Setting that kind of tone—violating an unwritten law and getting away with it—is moving the institution in a direction that it needs to go. Establishing that kind of caring and that kind of leadership is a long-term goal, and we need it among chairs and deans. But all of us can be leaders. All of us can step forward and help as senior professionals to move things in that direction.

Another long-term goal is to generate critical masses. Several members of our discussion group brought up this issue. Let me come back to AIChE and ACS meetings. If you begin to find even a fluctuation such that there are enough women in one place to have a conversation, you can make it happen. If you can force such an aggregation, then you will begin to feel more a part of things and be more comfortable in that environment. This won't happen overnight, but it is certainly a long-term goal.

Visibility is related to the shorter-term goal of networking. Our speakers raised this point this morning when we talked about getting national prizes, being elected to the National Academies. A long-term goal would be to move women chemists and chemical engineers toward more visibility in the community. Women will then be considered for these kinds of rewards.

Lastly, as a long-term strategy what we would look for is respect for personal priorities. I phrase this carefully because issues of women in the workplace often get boiled down to the issue of day care. But children need day care for only short periods in their lives, while other family-related problems continue. At my age I find increased responsibilities for taking care of sick friends or elderly relatives. These personal priorities—issues in your personal life that really matter and that have to be taken care of—more often fall to women than to men. This is a lifelong problem—it is not just a problem in your childbearing years—so respect for those kinds of personal priorities in the workplace matters for our careers.

We also talked about schemas and what I would call antidotes to schemas. One antidote might still be to use inclusive language. Another antidote is mentoring, in which we can help one another to see through these schemas and to sidestep some of this classification of women that we find handicapping.

A final antidote is development of our own self-esteem, which is a personal thing. A story in our group reminded us of the term that psychologists use, the "impostor syndrome." Many of us—more women than men—secretly think that we are impostors in the workplace. It's the feeling that one day we are going to figure out that we don't know what we are talking about and we will be done in—that we have gotten away with it so far, but only because others haven't figured us out—so eventually we are doomed. I think we women have many more problems with this kind of impostor syndrome, and development of self-esteem is something that we can use as an antidote to these schemas.

W. Sue Shafer, University of California, San Francisco: In our group we attempted to bring up both the positive and negative. I will give examples of schemas and of situations in which it would be important to have women as leaders.

One of the starkest stories described a department where there has never been a woman as department chair, but we also heard many stories about being the only woman in a particular situation. The clear conclusion was that numbers really do make a difference. Instead of trying to populate every place and every institution, maybe it would be better to identify just a few and bring up the numbers there. Even if everybody else were left to fend for themselves, it might be better having a few success stories to show what high-quality places resulted. Maybe then we would have built a better mousetrap that other people would want to emulate.

In another story, a young woman went to an institution where she joined an all-male laboratory group soon after the departure of another woman who had become pregnant and left the laboratory under emotional conditions. It appears that the men in her group felt that women achieved their ends by means that were inappropriate in the workplace. The narrator of this story was determined not to behave that way, but finally, after months of unsupportive treatment she asked, "What do you think you are doing anyhow?" and the answer was, "We are just treating you like one of the guys."

In a final story, Marjam Behar described her arrival at NIH to work in what was then called the Division of Research Grants. Her newly assigned female assistant came up to her and said, "I have never worked for a woman before, and I don't think I am going to like this." Needless to say, Marjam got her to see the light, and they worked well together over the years. I remember hearing that story when I was a fairly young manager, and it illustrates the point that sometimes women are their own worst enemies. If women who achieve positions of power don't treat employees in a caring manner but rather take on the worst characteristics of some men, the costs can be high. Unfortunately, everybody thinks you ought to know better because you are a woman.

We also discussed people being left out of the loop and what to do about it when this becomes a problem. In one person's experience, a man's unwillingness to include her in the loop was deemed her problem rather than his; she was counseled by male colleagues—who were otherwise very supportive of her—that in order to keep the peace in the group she was going to have to learn to live with it. I wonder whether a more creative solution to the problem might have been developed if her boss had been a woman.

Another theme that came up in several stories was the feeling among women that they have to do it all—and all at once—and that they have to do it all, but men don't. For example, such expectations affect how student evaluations are used as a part of the promotion and tenure process, although they are subject to the kind of schemas that benefit men more than women. There is some research on this (described in Deborah Tannen's book[1]), and one participant related that she intended to share it with

[1]Deborah Tannen, "The Argument Culture: Moving from Debate to Dialogue," Random House, 1998.

high-level officials in her institution. She was concerned that her chances of earning tenure might be compromised by the institution's use of what they incorrectly believed to be objective measures.

If we are talking about the consequences of not having women in leadership roles, one of the most important is that, until we have enough women in leadership positions for everybody to get used to it, women are very frequently not taken seriously when they assume leadership roles or positions of authority. We had a few stories to illustrate that.

We had two different views of industry that were expressed. One was that things look pretty good right now, that maybe most of the problems were solved for women in industry. On the other hand, we had a story about an R&D department in which every time a woman got pregnant she was transferred out of the department. This was told by someone who, while home recovering from the Caesarean birth of her child 2 weeks earlier and without adequate sleep, got a call from her boss offering her a transfer to the library. Needless to say she got up from the bed, put the kid down (presumably in somebody else's hands), and stalked in to work to raise the sensitivities of her male boss—who, by the way, did withdraw the offer of the transfer out of R&D.

I think we all know that leadership makes a difference. This is my bottom line. If leadership at the top is committed—regardless of whether the leader is a man or a woman, but preferably if women are seen to be high in the management team and if both men and women at the top are committed—we can make a difference, and we can set the tone, and we can change our environment. If we seriously want to invest in infrastructure to support women's careers in science, we can make a difference.

Geraldine Richmond, University of Oregon: Our group focused largely on the discussion of schemas and leadership; as an additional point we talked about mentoring, which you could also describe as championing or coaching. The discussion had as much to do with setting the framework for the workplace as it did about some soul-searching of where to go in the leadership area. The stories that we heard are disturbing and unfortunate, and the most disturbing part of it is that they don't seem to change—they are the same stories we have heard for a long time.

Another point that was brought up about schemas is that women must work very hard to put in a filter; they must decide which remarks may be negative toward them, which are just irritating, and which are harmful. Otherwise we would be buried in details.

A point that I found interesting in our discussions was that the schemas, or the atmosphere, tend to worsen with the success or perceived success of the woman. As one example, nurturing from colleagues is offered when a grant proposal is turned down. But when a person is very successful at getting grants and becomes competitive with the best people in the department, the nurturing may disappear, and then negative personality traits may appear when colleagues are uncomfortable with the woman's success. I must say that I don't think that applies necessarily just to women, but I think that it happens to women more often than to men.

We spent a lot of time talking about leadership and found that what results in someone being a leader in an institution largely depends upon the institution. Our group had representation from a variety of institutions (from government labs to small colleges, including women's colleges, to larger research universities and industry), and there were some differences particularly as our discussion focused on small colleges and particularly women's colleges. For the large research institutions and government laboratories, being accepted and respected as a leader largely depends on your research capabilities and how long you stay competitive or active in research; there is little correlation with actual leadership skills. Your ability to be effective up the ladder in order to bring change to your unit often depends on the reputation that you bring external to the institution. Consequently, if women are to hold leadership positions in some of these institutions they must maintain their research credibility. I'm not sure that

that is a message that we are sending down the line to the troops. It can become a very important factor in achieving career recognition—becoming a department head or being elected to the Academy. I think these issues are important to address on a national level. Achieving the necessary research success—in government laboratories as well as academic institutions—is dependent on maintaining research funding, credibility, and visibility.

There was a discussion of the "male" method of running a unit or a department, which again was a perceived extreme. But most of us see males as department heads, and the extreme is the autocratic rule of a linear thinker, someone who stays inside the box with regard to thinking. I don't think any of us would necessarily say that is the best model. The other extreme is that of the "female" model. One of our group members observed that this other extreme is not necessarily productive either: faculty meetings or unit meetings that are just a free-flow discussion, soliciting everybody's opinions, and letting everyone feel nurtured, wanted, and needed—without necessarily coming to any final decision.

The question is, in that spectrum of leadership, where are women most comfortable? Where do women feel they can be effective and still retain the qualities that they perceive are important? To be a leader do you really have to become an individual that you dislike, or can you be an effective leader and still retain the values that you respect? Where is the happy medium between linear thinker and what was called circular thinker?

On the issue of mentoring and championing it was emphasized that mentoring requires work and training for both the mentor or the championer and the protégée. The mentor must understand how important that role is—it is not just having lunch with the person once in a while—to actually help open doors and promote that person. At the same time, the protégée must be willing to take criticism, to say, "Well, that's all very nice, but now tell me what really is wrong," and then move forward.

I am struck by the lack of progress we are making in how women are perceived or evaluated in their jobs—whether the evaluation is in terms of our research, our effectiveness as department heads, our leadership roles, or our efforts as mentors or champions. We are still searching for how best to be a leader, how best to be a mentor, how best to be a champion, and how to take on these roles. There are so few women in leadership roles that we don't have enough examples of effective leadership among women in the chemical sciences to know who are the appropriate role models. We need more women as leaders, not only as role models, but also as champions and advocates for other women coming up the line.

Lou Ann Heimbrook, Lucent Technologies, Bell Laboratories: The stories discussed by our group all fall into three areas. One was related to the consequences of supporting specific individuals, a category into which all of us fall at some point in our careers. The particular situation that was brought up involved the only female faculty member on a search committee during a search to fill a faculty position. A female candidate was hired, but she left the university after only 6 months. To whom did the other members of the original search committee turn and ask, Why did that person leave? What is the issue? To their female colleague on the search committee, of course. We need to continue to support women candidates, but all the stories showed the possible consequences.

A second theme in the stories was the frequent exclusion of women from decision making, in both academics and industry. A case in point was the story of a particular department that was diverse in both racial and gender characteristics. There were two females in this particular department, one tenured and one not; the faculty also included a group that basically knew they were in the retirement phase but still wanted to decide the future of the department. As they held meetings and activities about planning the future of the department, some of the faculty—including the two females—were excluded. Perhaps this was just the tenured versus non-tenured? Unfortunately, no! One of the women was a tenured professor. I think it all comes down to what we heard earlier, which is the question of critical mass.

So what were the schemas that caused the problem? I alluded to this before. There was the issue of separation, in which men held separate meetings to make decisions. If you are not included, it is tough to make a difference. A second problem is, "When in doubt, blame the woman." If another female comes into a department, whether in academics or industry, and doesn't succeed or simply doesn't stay, there is a retention issue that acquires an immediate association with all women in the department. The third concern is the titleless woman, an issue that appeared in several of the stories that we heard.

What are the consequences of these stories? They leave us with a discomfort level, and we all know what we do when we become sufficiently uncomfortable. We don't show up. We opt out, and we can give a lot of reasons for opting out; but if enough of us opt out we will never get to the critical mass. Another consequence is the immediacy of need for women in higher positions. I call this the need for speed. The statistics that have been presented here show 10- and 20-year gaps that consistently seem to look the same. One of the consequences is we are running out of time. There is a need for speed in many of the activities that we need to drive if we are going to get critical mass and if we are going to make a difference.

What are some potential solutions to these problems? First and foremost is continuous attention—by both individuals and institutions. We need accountability and responsibility, and we need to hold other people accountable and responsible for the actions that we are trying to implement. And we need to use positive and not negative reinforcement.

Frankie K. Wood-Black, Phillips Petroleum: Among the schemas that you all notice is that the lapel microphones are designed for men in business suits with pockets. Women's attire is not necessarily conducive to business suits and pockets. We've been seeing different solutions to this during the conference.

Our discussion group was smaller than the others, with a good mix of people from industry, university, even some representatives from government.

Two of the industries in the group were my company (Phillips Petroleum) and Bell Labs (formerly AT&T Bell Labs, now Lucent Technologies), both of which have been recognized as being positive places for women to be. Phillips Petroleum has two women on its board of directors, which is unusual for a petroleum company, and Bell Labs has made a huge effort over the last 10 years to incorporate women and help women make progress up into the management ranks. We sat in back and said, Let's look at how the workplace has changed in the last 10 years. We looked at the efforts other companies were making and what different organizations were doing and we concluded, Oh, man, we were successful. We now have women vice presidents. We now have two women on the board. We are successful. This is great.

Wait a second! What happens when those new executives retire or move on to other opportunities? Where is that new tier? Can you think of a woman who is in the progression or the chain of promotion who is going to be the next vice president or the next board member? You can think of a whole bunch of entry-level women, but you cannot necessarily think of women who are in middle management on the way up.

So, were we successful? I don't think so. Something is happening in the mid-tier. Organizations are not dealing with this mid-tier very well. What is happening and what is going on? The organizations have been working really well to get women into positions of leadership. The question is, How do we keep them there and keep them in the chain?

It worked out really well when our breakout group began to talk about the mid-tier, because then we just kind of rolled right into the different schemas. By coincidence, our group included Mary Jordan, the ACS statistician who deals with all the career surveys. She is going home with lots of opportunities to follow up on these discussions. There are questions about women in mid-career for which we did not have the answers. Where are they? Where are they located?

One of the big things we are seeing is the change in the workplace in big chemistry. If you went to work in chemistry in the past, you went to a place such as Dow, DuPont, Phillips, Mobil, or Exxon. Wait a second. I just gave you a whole bunch of names of companies that either aren't there anymore or have merged with other companies. But now you can open any journal and see 50 small pharmaceutical start-ups, especially chemical start-ups, little companies. So, it is big chemistry versus small chemistry. What's happening there? Is that changing our workplace dynamics?

Those questions brought us to the subject of group dynamics, and we discussed the effect a woman has in a meeting. Is she heard? Not necessarily. Is she looked at when she is talking? Some people talk around her. Is she excluded? The critical issue is group dynamics, where communication styles also come into play. Is the woman's behavior looked at positively or negatively? If a woman interrupts she is being aggressive, she is viewed negatively; but if she is not interrupting, she is not being heard. So this is a complex situation, and power positions don't necessarily correlate with reactions.

A woman in a team-leader or managerial role is not necessarily accepted as a leader. A junior person in a hierarchy can still tell you, "That is just your opinion. You don't have any basis for that." A woman in a position of power doesn't necessarily elicit the respect that a male counterpart would receive in the same position.

You also heard in several of the other groups about the problem of women's self-image. When women succeed, they sometimes consider themselves impostors. "Oh, I was lucky. I got a break." "This is really good. I got a break." "If I don't win, it's not because I tried and I did everything I could, but there was somebody else out there better than I was." " It was because I didn't work hard enough." So it's *lucky* versus *lazy*—We are always looking at it as our being lucky or lazy. It is not talent based. Men, on the other hand, say, "That guy—that professor—was just a jerk. He didn't recognize my talent." Men do not necessarily blame themselves. So there is a self-image issue.

That discussion brought us to the consequences—of not having people in that mid-tier or not having people in the hierarchies. Everyone sees it as paying the cost to get there.

We heard stories this morning about how women who had advanced in hierarchies typically did not have children. They had to make a choice between career and family, and that is seen as paying a cost. It may not have been a cost to that person who made that choice, but it is perceived as a cost to someone who is in the lower tiers: "I want to have both a career and family," or "I don't necessarily want to work 60 hours a week in the laboratory. I want to be able to do other things."

Those are costs—or at least perceived costs—and generally those costs have a negative image. We have only a few role models, and we ask, "How did that person do it?" From my experience there are role models out there. I know a woman with nine children—a successful dual-career family, and nine children. How did she do it? I couldn't do it. I have two children, and we have a little problem with that. Seven more children? No, I don't think so.

People see the sacrifices, and the lack of role models causes students to look and say, "That person spends 80 hours in the laboratory, writes three grant proposals a week," and so on. They say to each other, "I don't want to pay those kinds of costs!" because they don't have other role models to illustrate other solutions to the problem. There are multiple solutions here to the problem, but there are not enough women in the middle and upper tiers to demonstrate the solutions. We need to show our younger colleagues that there is more than one way to skin that cat.

Career path choices: these are our trailblazer issues. You need a trailblazer to show the company and to show younger women that there are options. That is why we must have creative thinking so that someone can say, "Look, you want this, I want this. This is how we can get a win/win situation." We still tend to focus on the costs—the self-image issues and the career path choices—as demotivating forces. We are getting hit in the stomach all the time, and—you've all watched boxing, a very masculine type of thing—it's always the stomach punches that get the guy in the end, because they're one after another, and they weaken. These are the demotivating forces that we need to overcome.

DISCUSSION

Marylee Southard, University of Kansas: Someone asked for more data from the statistician in their group. It may be extremely hard to track, but where are the engineers and chemists going who go directly out into the workplace with their B.S. degrees? We have been producing between one-quarter and one-third women for 15 years or so, but we don't see those numbers out in the workplace in industry and government. It is very difficult to track those. Is there a way to go about it other than asking individual alumni groups to get out their address files? I ask because our advisory board has addressed this as well, and we cannot find our own alumni in the workplace.

Janet G. Osteryoung, National Science Foundation: Could I just make one comment about this? I think people should be clear on the fact that the only reason NSF has a reasonable database for Ph.D.s is that somehow they have conned everybody into thinking that nobody can receive a Ph.D. unless they fill out this form and send it in. But this may be an insurmountable problem that you are addressing.

Mary Welsh Jordan, American Chemical Society: The ACS data look very good. We have survey responses from 50 percent of the Ph.D.s and 40 percent of the new graduates with bachelor's and master's degrees. That is better than most surveys do, and compared with the national data that we have, it is pretty well right on. So I think they are considered very sound data, even by NSF and other organizations.

Janet Osteryoung: I think we all owe ACS a debt for persisting in collecting these data. There is no question in my mind that they are the best that exist.

Robert L. Lichter, Camille & Henry Dreyfus Foundation: I just want to elaborate a little bit on some of the discussion in our breakout group. The notion of immediacy of role models has to do with not only the speed of getting people into the higher positions. There is also a need for those in higher positions to have a more immediate relationship with people in the lower positions. It is fine to have women who are presidents of academic institutions and CEOs of corporations, but at the entry levels it is important to have women who are just above the women who are coming in as well. That is the other aspect of immediacy.

Another notion that was discussed in our breakout group was the idea of creating pathways of training. A number of corporations are preparing employees for management positions, but this seems woefully absent in academic institutions—indeed, it seems to be antithetical to the notion of an academic institution.

5

An Oddity No Longer: Women Scientists in Industry

L. Shannon Davis
Solutia, Inc.

I want to thank Dr. Janet Osteryoung and her colleagues for putting together this workshop; it is a timely, very relevant subject. I also want to thank my employer, Solutia, Inc., for allowing me to speak today.

I'm going to present a perspective on women scientists in industry. "An Oddity No Longer" is an apt title for this discussion, as the statistics and some of the anecdotal evidence I'll be presenting amply demonstrate. This will be a balanced picture, because by no means does the industrial world have the perfect solutions to attracting, retaining, and promoting outstanding women scientists. Nor is what I'm going to say true of all segments of the chemical industry—including my own company. I've culled some success stories, as well as some tales of woe, from leading companies in the chemical industry in hopes of gaining some insights into what's gotten us as far as we've come and maybe to shed some light on how to continue to make progress. I'll also raise questions that I don't have answers for, which I hope will provoke some healthy discussions about what is working, what is not working, and some potential solutions to keep talented people contributing in our workforce.

I'll begin with an overview of some statistics. I suspect that by the end of this workshop we all will know these numbers by heart, as I'm sure all of us are quoting them from the same or similar sources.

First of all, we'll talk about the pipeline. Where are the women pursuing science and engineering degrees? What does the profile look like and how does it affect our hiring and promotion practices? We'll talk some about the ramifications of the pipeline as well—changes in today's workforce, for example. Then we'll segue into trends for today's workplace. These trends impact what the workforce expects and what you have to do to retain the best and the brightest for your company, your laboratories, and your lecture halls. We'll look at some programs that have been successful for industry and some that haven't perhaps been as successful as we'd like. And we'll talk about possible trends that will exacerbate the exodus of women to other careers. I'll close with some suggestions on how to capitalize on this information and some critical questions to ask about your workplace and your organization that I hope will leave us with some food for thought for tomorrow's sessions on critical factors for success.

The data in Figure 5.1 show the number of chemical engineering graduates by degree and by year.

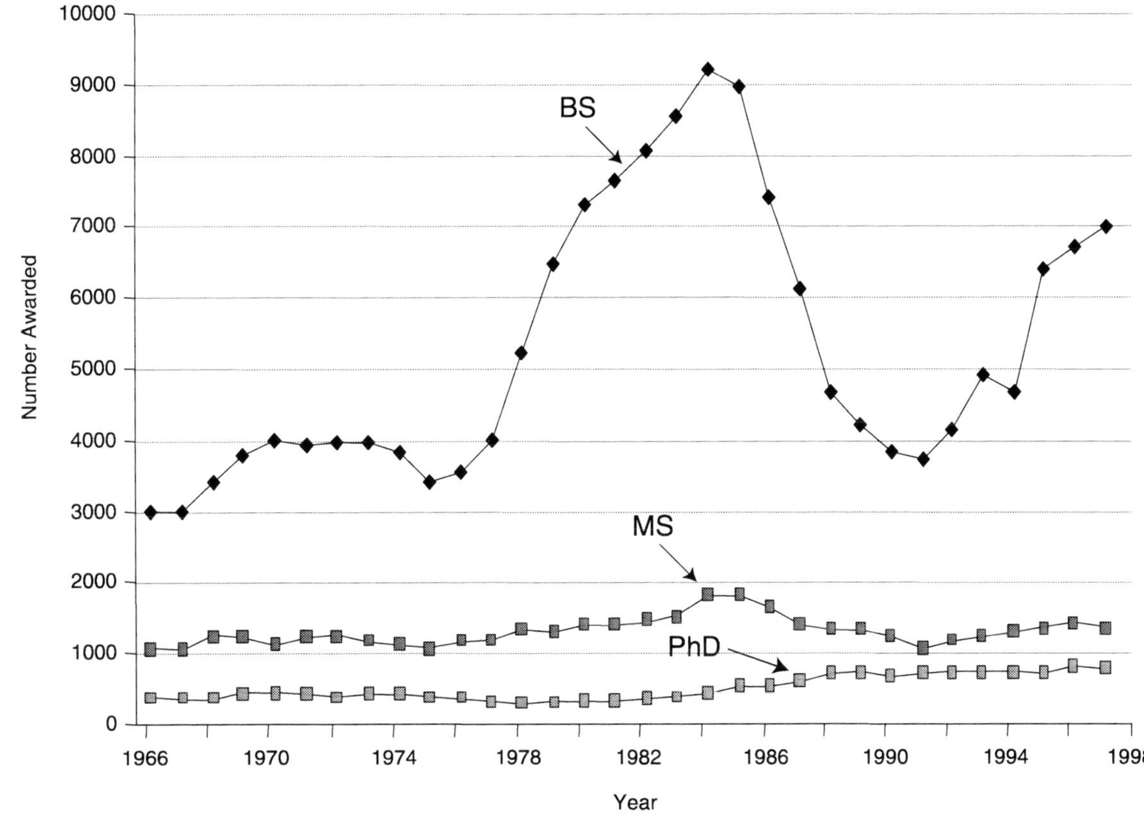

FIGURE 5.1 Chemical engineering degrees awarded in the United States, 1966 to 1997. SOURCE: National Science Foundation, Division of Science Resources Studies, *Science and Engineering Degrees: 1966-97,* NSF 00-310 (Author, Susan T. Hill), NSF, Arlington, VA, February 2000.

I've broken out the data to show just chemical engineers, as this is the principal discipline in the science and engineering fields that moves into the chemical industry. Note that the slowdown in the production of engineers in the early 1990s in response to the declining economic conditions of those years is in a full-swing reversal.

Figure 5.2 shows similar data for chemistry graduates, again by degree. We see a similar impact of the mini-recession in the early 1990s, followed again by a nice recovery; there are graduates out there from the disciplines relevant to most of the chemical industry. I deliberately selected these two segments for particular focus as they make up the majority of the technical hiring pool that we in industry draw from.

Figure 5.3 breaks out degrees by level and by gender. Starting in the 1980s, you can discern a relatively steady increase in the number of women choosing chemical engineering degrees at all levels. A similar trend can be seen in Figure 5.4 for chemistry degrees, although it started a bit earlier there.

Perhaps a slightly clearer indicator of the increasing number of females receiving science and engineering degrees can be seen in Figure 5.5, which shows women as a percentage of the total chemical engineering degrees granted. The numbers steadily increase over the years, so that today's graduating classes of chemical engineers are around 30 to 35 percent female. The dark horizontal line on Figure 5.5

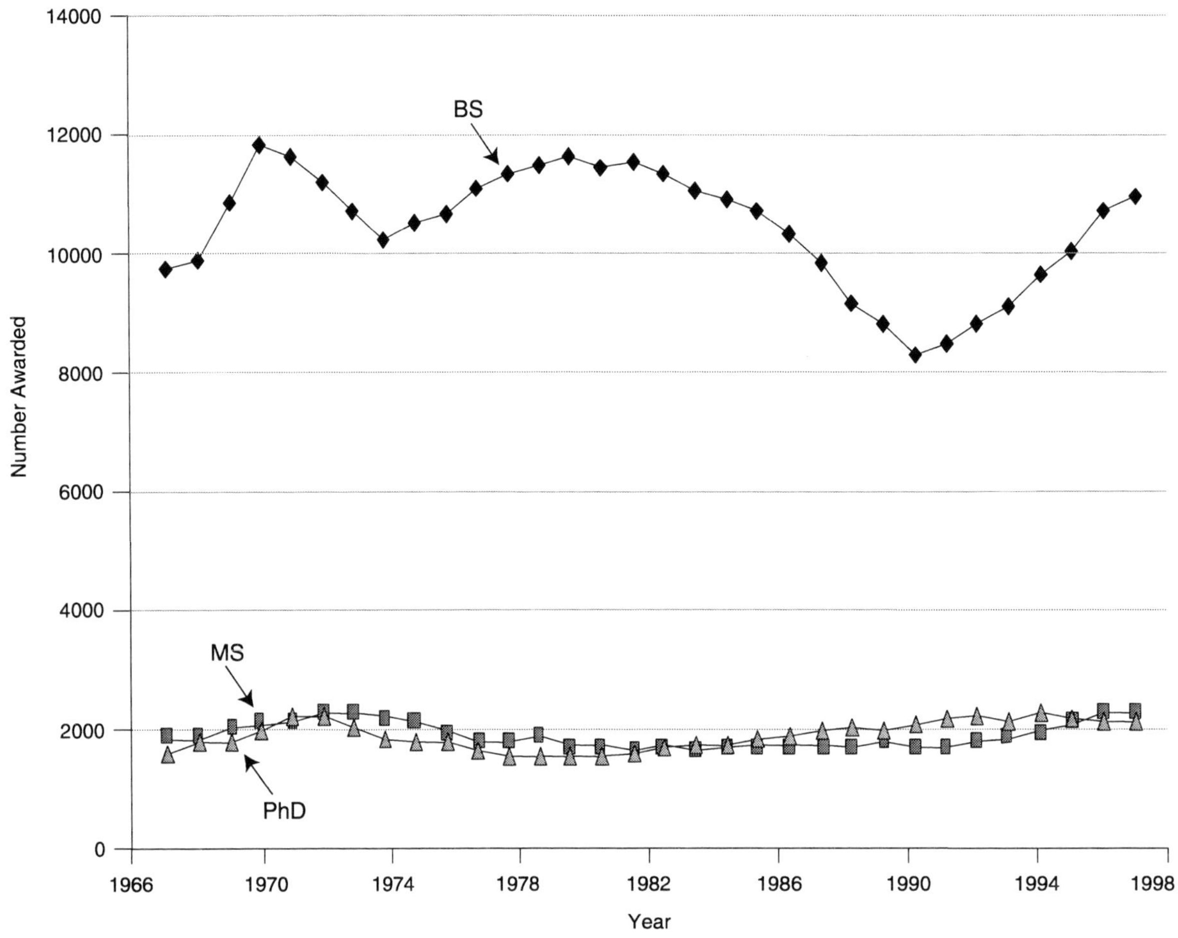

FIGURE 5.2 Chemistry degrees awarded in the United States, 1966 to 1997. SOURCE: National Science Foundation, Division of Science Resources Studies, *Science and Engineering Degrees: 1966-97,* NSF 00-310 (Author, Susan T. Hill), NSF, Arlington, VA, February 2000.

is an arbitrary distinction that I chose in order to facilitate discussion later on about "the pipeline theory." At 20 percent of degrees granted, women make up a large population. At this point, the entry-level pool of candidates is fully supplied with qualified candidates. Note that for industry, where the majority of chemical engineering hires are at the B.S. level, we reached this arbitrary distinction point in 1982—almost 20 years ago.

The share of chemistry degrees earned by women shows a similar trend (Figure 5.6). We see a continuing increase in the number of women receiving chemistry degrees at all degree levels. In this case, however, we reach the arbitrary 20 percent mark much earlier. For B.S. chemists, who would typically be hired into industry at technician or junior professional levels, we hit 20 percent in 1974; for Ph.D.s, who are typically the degree level most often hired by industry, we reached that level in 1986.

My proposed hypothesis is that these data indicate rather strongly that the pipeline is full and has been for at least 20 years. Women are receiving increasing proportions of the science degrees awarded and have been for a while. Thus, a logical conclusion is that the hiring pool of talent today contains enough women, and has for the past 15 to 20 years. Problems with finding qualified female candidates

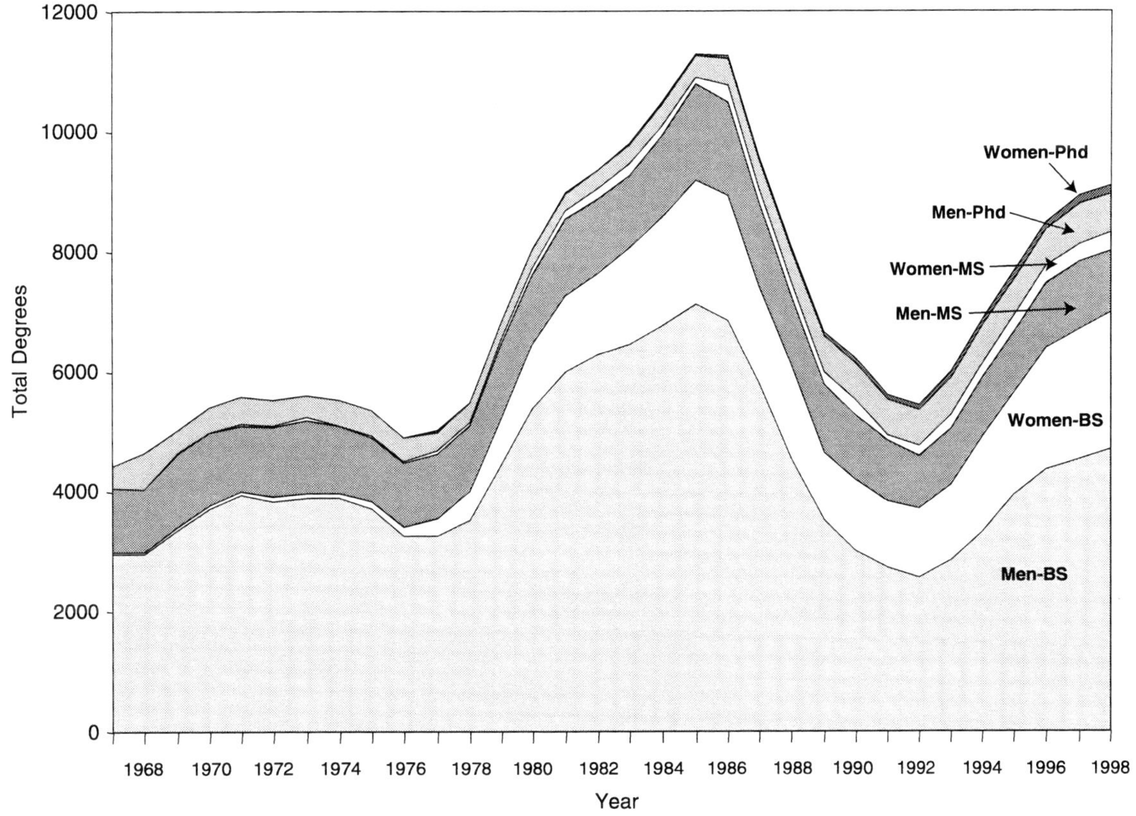

FIGURE 5.3 Chemical engineering degrees awarded in the United States by gender, 1966 to 1997. SOURCE: National Science Foundation, Division of Science Resources Studies, *Science and Engineering Degrees: 1966-97,* NSF 00-310 (Author, Susan T. Hill), NSF, Arlington, VA, February 2000.

for jobs—particularly entry-level jobs—should be alleviated with women accounting for over 20 percent of the degree holders.

If this hypothesis is true, then a few other logical conclusions also follow.

If we've had a full pool of talent for 20 years, then a snapshot survey should show women at all levels of industrial jobs today. Even with some attrition, we should see women attaining the upper echelons of career paths, whether that is in management or in the technical track.

So where are they today? In industry, these numbers are very hard to come by. Madeleine Jacobs, editor in chief of *C&EN*, took the approach of perusing the annual reports of the top U.S. chemical companies. In 1997, only 4 percent of the upper management listed in the annual reports of the top 19 U.S. chemical producers were women, with the majority of these holding staff titles. Numbers on technical staff are virtually impossible to find. A 1998 study of Fortune 500 corporations by Catalyst, the nonprofit organization that aims at maximizing opportunities for women in management and executive positions, showed that fewer than 4 percent of the chairman, CEO, and VP titles were held by women. Only 11.2 percent of corporate officers were women, and they projected very slow growth rates (13 percent in 2000, 17 percent in 2005).

These are some of the demographic data. Now I'm going to discuss trends. I'll present some general workforce statistics and then discuss hiring, recruiting, and retention from an industrial perspective.

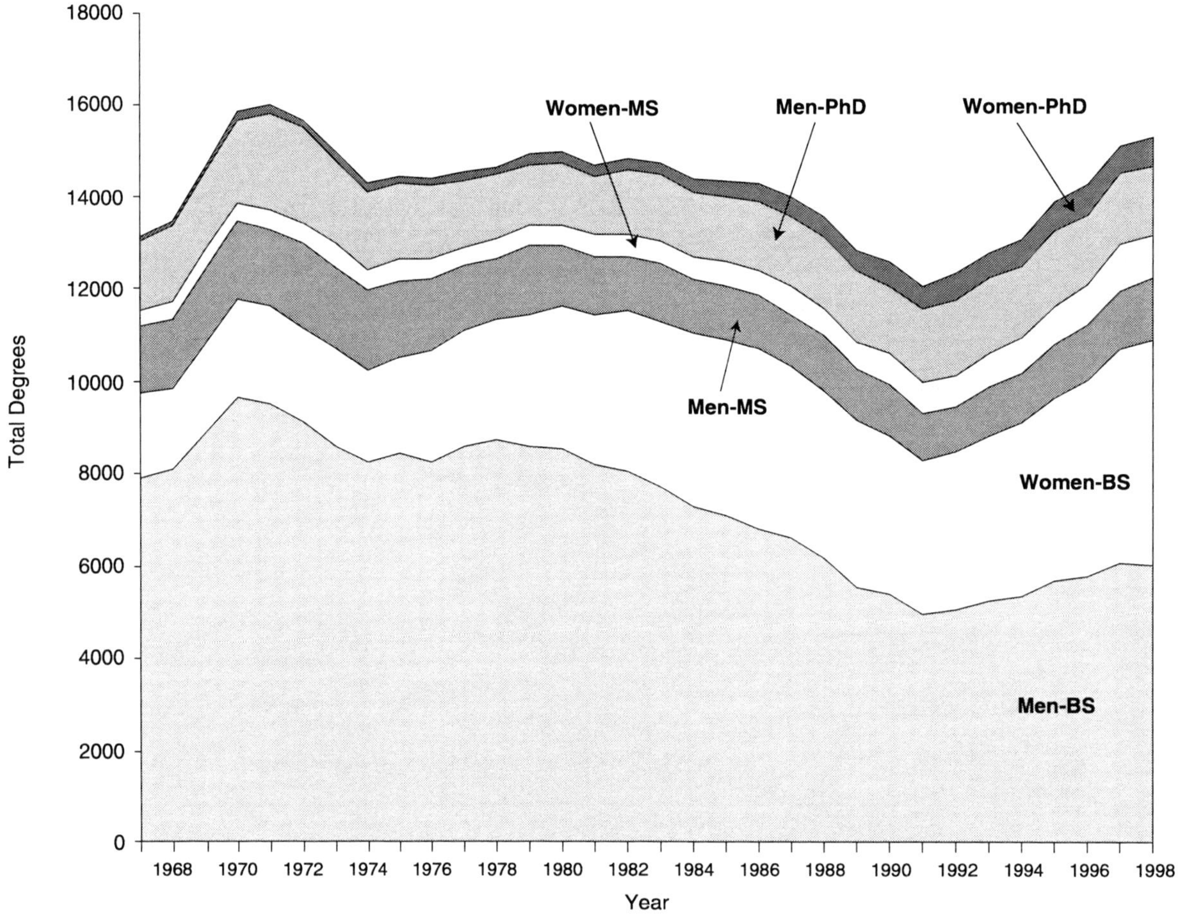

FIGURE 5.4 Chemistry degrees awarded in the United States by gender, 1966 to 1997. SOURCE: National Science Foundation, Division of Science Resources Studies, *Science and Engineering Degrees: 1966-97,* NSF 00-310 (Author, Susan T. Hill), NSF, Arlington, VA, February 2000.

I'll also review some of the policies and programs that are changing to meet the changing demands of the workforce today.

Several trends worthy of note emerged from the NSF science and engineering indicators study in 1998; these are summarized in Box 5.1. First of all, the science and engineering professions enjoy extremely low unemployment rates—about one-half of the national average today—and the projected demand for scientists and engineers over the period of the study (through 2006) is expected to increase substantially faster than the national average. Also, current trends in degrees granted and numbers of people employed in the science and engineering fields indicate that the shortfall of trained scientists and engineers that was so highly touted several years ago is not going to occur. While record numbers of science and engineering professionals will begin retiring as the baby boomers hit retirement age, there seem to be sufficient numbers of people entering these fields to replace them, so the age equation seems to be pretty much in balance. A key issue here is the more forward-looking indicator of the number of people choosing to major in science and technology fields. If fewer and fewer people are choosing to enter these fields, the supply and demand equation may not be as much in balance as we think.

A significant trend in the workplace has been the rise in the number of dual-career couples seeking

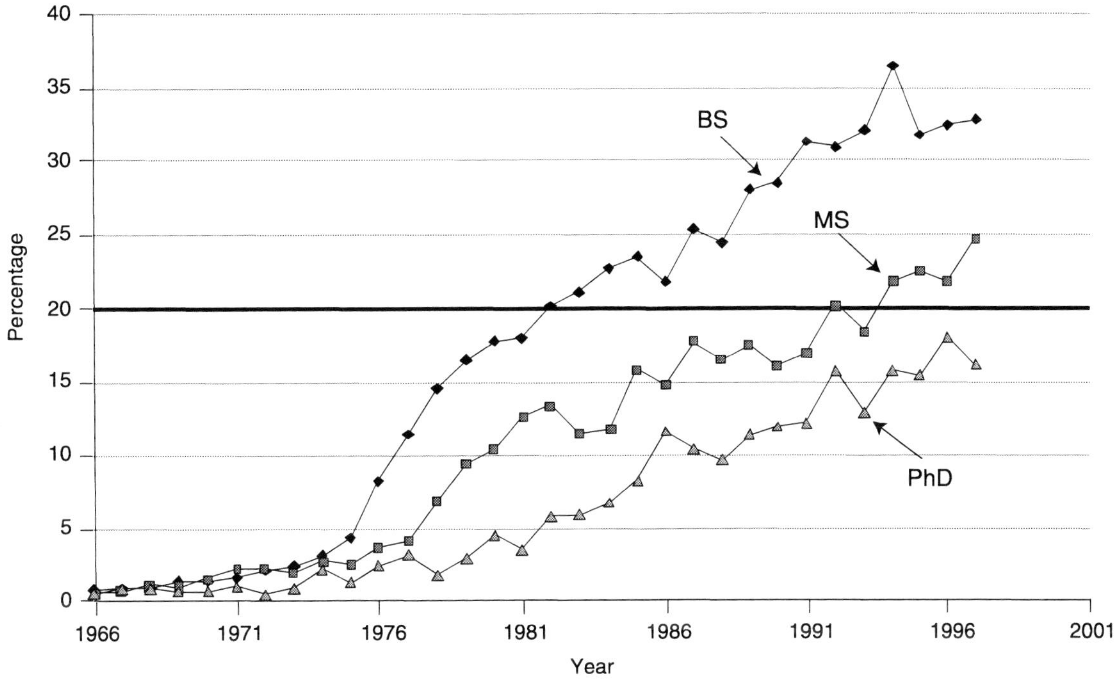

FIGURE 5.5 Chemical engineering degrees awarded in the United States, women as a percentage of the total, 1966 to 1997. SOURCE: National Science Foundation, Division of Science Resources Studies, *Science and Engineering Degrees: 1966-97,* NSF 00-310 (Author, Susan T. Hill), NSF, Arlington, VA, February 2000.

jobs, as shown in Box 5.2. Since 1960, the fraction of dual-career couples has risen three times, to over 86 percent of the workforce. More than half of marriages in 1996 were dual-income couples, and they made up 45 percent of the workforce. These changes have driven changes in many other areas as well—benefits policies and growth industries such as day care, Internet shopping, and the like.

As we've already discussed, there is a large and diverse pool of talented, qualified candidates. In industry, recruiting for both women and minorities is no longer haphazard. Special efforts are made to recruit talented individuals in both categories—by hosting pizza parties for the women in engineering

BOX 5.1
The Science and Engineering Workforce

- Science and engineering enjoy low unemployment rates (~2%).
- Demand for science and engineering occupations is expected to increase over the next 10 years at three times the national average.
- Expect an increase in trained scientists and engineers for some time.

SOURCE: National Science Foundation, National Science Board, *Science and Engineering Indicators 1998*, NSB 98-1, NSF, Arlington, VA, 1998.

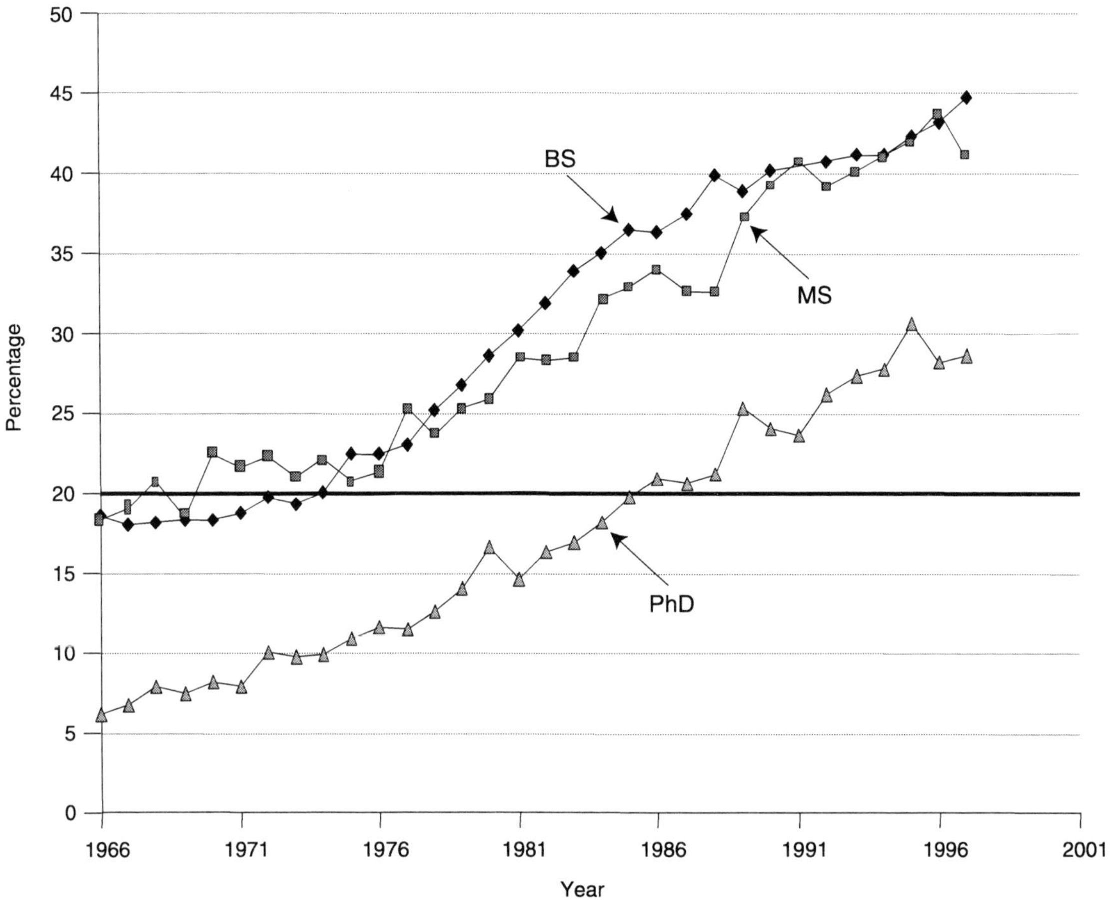

FIGURE 5.6 Chemistry degrees awarded in the United States, 1966 to 1997, women as a percentage of the total. SOURCE: National Science Foundation, Division of Science Resources Studies, *Science and Engineering Degrees: 1966-97,* NSF 00-310 (Author, Susan T. Hill), NSF, Arlington, VA, February 2000.

societies on campus, for example. To raise candidates' awareness of a company, some businesses fund special events, programs, or scholarships. And some industrial recruiting efforts target specific schools. Industry has also recognized the need for support structures, with new-hire orientation programs, communications classes, and, in some cases, special programs for the needs of dual-career couples.

Our cafeteria-style benefits have also changed substantially in response to these changing demographics. In the area of retention, three things stand out. The first is salaries and benefits and pay equity measures. The second is career paths for science and engineering graduates in industry. The third critical factor in retention is often the policies and programs available in-house. A fourth, more nebulous factor in retention is "atmosphere," which is hard to measure, quantify, and affect. Atmosphere often falls in the category of things "you know them when you see them."

Figure 5.7 presents a graph of salary, by gender and degree level, from last year's ACS salary survey. These data are principally for chemistry degrees and are only for industry. A couple of interesting points emerge. There are no data for women in the 30-year-plus brackets. Having seen the historical degree data, this is not a surprise, as it reflects the virtual absence of women with degrees in

BOX 5.2
Dual-Career Couples

- Two-wage couples

1960	25%
1980	50%
1990	86%

- Dual-career couples made up 60 percent of marriages in 1996.
- Dual-career couples were 45 percent of the workforce in 1996.

SOURCE: Catalyst, "Two Careers, One Marriage," New York, 1997.

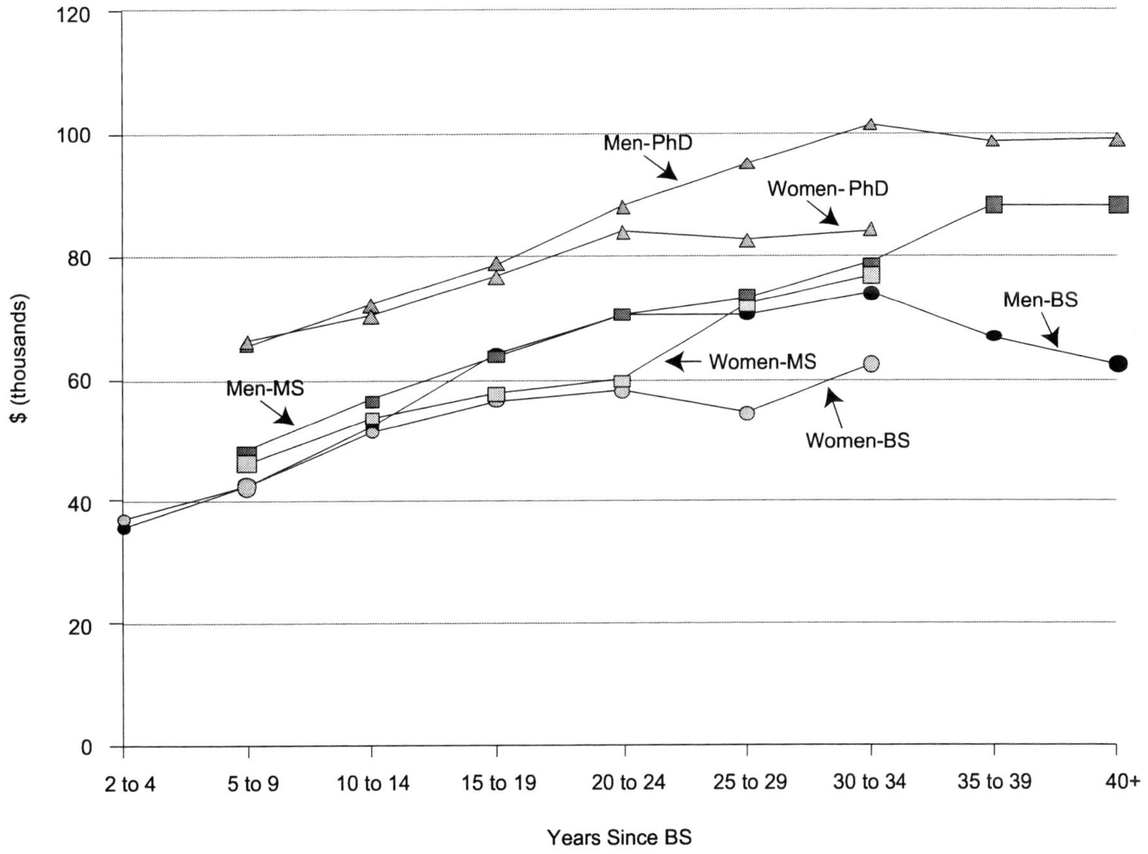

FIGURE 5.7 Industrial salaries by gender. SOURCE: Michael Heylin, "Salary and Employment Survey," *Chemical & Engineering News,* August 2, 1999, pp. 28-39.

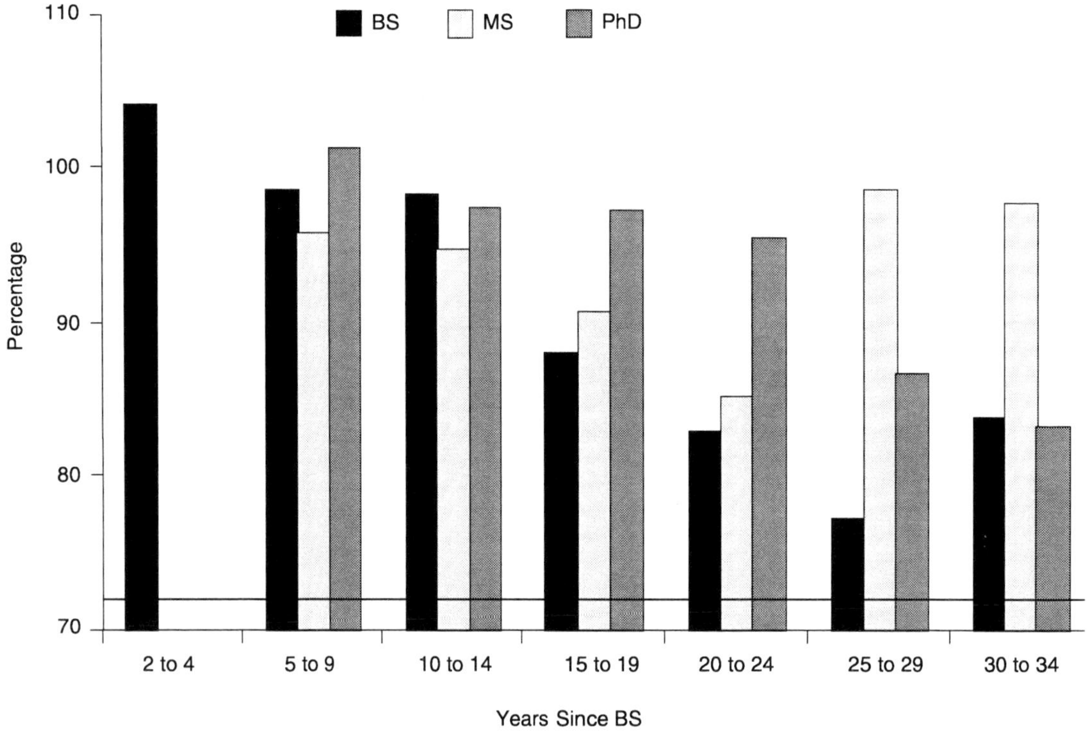

FIGURE 5.8 Industrial salaries, women as a percentage of men. SOURCE: Michael Heylin, "Salary and Employment Survey," *Chemical & Engineering News,* August 2, 1999, pp. 28-39.

that time frame. Overall, the trends in the early years are very positive, with women holding their own in terms of salary. The salary divergence gradually creeps in as years of experience increase, and it gets quite noticeable in every degree category about 15 to 20 years past the B.S. degree. This is very interesting, given the question I raised earlier about pipelines and "where are they?"

Figure 5.8 presents these same data from a different perspective, showing women's salaries as a percentage of men's. There is a definite drop as years of experience increase, and there is a worrisome increase in the gap as the years past B.S. increase. Several factors have been posited to explain this gap—women leaving the workforce for family reasons, and so on. This is also closely correlated with the numbers of degrees granted. The increased disparity didn't start until about the 20-year mark. Note the horizontal line at about 72 percent; this is the national average for the United States.

Let me make some comments on pay equity. It's hard to believe that we still have to talk about this subject in the year 2000, but there is still a small wage gap in industrial salaries, although in the early years it is very small. In the ACS survey, there are insufficient data at high levels of experience to report—few women with 30-plus years of work history.

Some interesting questions arise in these studies as well. Again at that 15-year mark the gap increases, as we saw also in Figure 5.7. What is so significant about this 15-year mark? A similar study by the NSF in 1999 found a definite correlation between years of experience and salary. Their data, covering a wider range of science and engineering salaries and including academicians, showed a pay equity gap of about 87 cents to the dollar. They also showed a big difference in years of experience.

When that particular variable was accounted for in the regressions, the wage gap closed to 97 percent. Women, in this study, had on the average 5 years less experience than men, and this experience level difference was key in the wage gap that was found. However, they also noted the absence of "older" women in the engineering workforce and that women with engineering degrees more often were employed outside their field. Again, why?

One of the keys to retention is having mechanisms in place to allow employees to grow, stretch, and learn. In industry, a technical person will typically follow one of two major career paths—either staying technical for the duration of their career or moving into management. In many places, the technical track is referred to as the "fellow" program, after the IBM and, now, the Bell Labs programs begun in the 1970s. This dual-track career ladder is intended to ensure that people who choose to remain at the bench for their careers are compensated and have the ability to achieve titular levels equivalent to those of management and commercial-track people. Some people move into the commercial side of the business and follow a commercial track. This is one key area where dropouts occur, as these people typically no longer respond to surveys of the science and engineering workforce.

One of the advantages of these multiple tracks is the flexibility in choices, particularly for dual-career couples. These ladders enable people to take on new challenges, in many cases without having to relocate families, and grow within the same company or organization. For dual-career couples, this flexibility allows both partners to enjoy more fully their job or career choices, and while sacrifices are still in many cases required, the degree of sacrifice is somewhat less.

A tried and true tenet of management is that "what gets measured, gets done." This premise is the foundation for most of our management programs in industry and for many other programs, such as quality assurance. If you want to increase the proportions of women and minorities in your organization, then you must begin to measure the numbers. Where are they? Where are they going? Why do they leave? These are all key questions that should be monitored. Both formal programs and benefits and informal programs are key to providing the support structure necessary to create the right environment to retain *all* of your best and brightest, not just the women. This is a key area that defines that nebulous "atmosphere" quality in a more formal fashion.

In the 1980s there was a big push for diversity training, sensitivity training, cultural training—the movement had many descriptors. These programs were generated in response to a growing recognition that different groups of people think about the same things in very different ways. Our stereotypes determine how we filter and interpret information, and our cultural upbringing subconsciously influences those filters. These sessions were intended to instill in participants an appreciation for other perspectives and to begin to highlight some of the more subtle forms of discrimination that exist, to bring to the surface some of those filters that resulted in the famous glass ceilings, and to become more aware of subtle barriers to success that exist in organizations, work groups, and individuals.

While in most cases this training was well received, the increased focus at the same time on both diversity issues and sexual harassment had some negative ramifications in the industrial workplace. Some more subtle biases and barriers began to crop up as backlash—cases in which men in an organization were reluctant to work with women because they weren't sure how their actions would be interpreted, because they didn't want to deal with potential harassment issues.

"Early ID" is a management technique that is routinely used in industry and other places to identify good, promising candidates for high-level jobs early in their career. These people are typically high-potential candidates for a large number of jobs within the organization and so are groomed, with more attention given to their career paths by management. This attention typically takes the form of high-visibility assignments—a key criterion for promotion—good skill-building opportunities and other such grooming opportunities. Exposure to the top brass is critical. These people are also mentored and coached to fully develop their potential. In many organizations, part of the measuring includes a dedicated "people planning" process, in which personnel are discussed in terms of their career potential. Specifically discussing women and minorities who are interested in technical or management-track careers is a key component of this planning process.

Mentoring is both a formal and an informal process. Many organizations—General Electric, Procter & Gamble, and Corning—have formal mentoring programs, which in many cases go hand in hand with their diversity training—part of broadening cultural and business perspectives. Formal programs typically involve pairing up a senior with a more junior person, to show them the ropes of the organization—mores, standards, cultural norms. This person can also be a development coach, offering suggestions on how best to pitch that great new idea or who in the organization would be most receptive to it. Also, mentors can help with career choices and with internal networking. Mentoring is a critical component of the atmosphere. Good places have it.

More informal networking happens at professional society meetings, workshops like this, and with organizations such as the Women Chemists Committee of the American Chemical Society or the international American Women in Science. It also happens informally in organizations with social or work groups. And the advent of the Internet has expanded greatly the opportunities for networking, with online career coaches, chat rooms, and the like.

The changing workforce demographics has dictated sweeping changes in both policies and benefits for the industrial workforce. The Family Medical Leave Act is much more frequently used for new family additions—and by both parents, not just the mother—since it was introduced into law in 1993. Twenty million workers have used the law since it was passed; 58 percent were women, and most use was for personal illness. A newer change, which has been enabled by the rapid growth of technology, has been alternative work arrangements. These can include part-time work, job sharing, telecommuting, flex time, alternative shift schedules, and other creative ways to allow people to manage more effectively their work lives and their personal lives. The flood of dual-career couples has also eliminated nepotism rules in industry. While there are still some common-sense restrictions on husbands and wives employed by the same company, most of the—by today's terms outlandish—rules of nepotism have long since been left out of corporate policy manuals. The switch to cafeteria-style benefits—which include before-tax contributions to day care, dependent care, elder care, and so on—helps in customizing what families today need and are willing to pay for. Some issues left to deal with here revolve principally around bench work. Telecommuting works very well for some job functions, but hands-on laboratory research is not one of them. Helping career scientists use some of these tools is a significant challenge to today's technology management.

I've highlighted some of the programs that seem to be working in industry. Now let's talk about some things that aren't working. First, let's talk about the dropouts I referred to earlier. Many people with science and engineering degrees who follow industrial career paths end up in jobs that do not show up on science and engineering surveys. People who pursue commercial business management tend not to classify themselves as technologists and so are lost to ACS salary surveys and the like. More and more people are achieving some semblance of their goals only to find out that "corporate America" isn't really what they want to do with their lives, and so they drop out to pursue other careers. Entrepreneurs and small business start-ups with women at the helm have skyrocketed in the last decade. Many of our talented women are choosing to make their own rules by running their own businesses instead of bending to corporate America's dictates. In many cases of balancing work and family issues, family is taking precedence, and we are beginning to see an increase in the phenomenon of re-entry—women who chose to stay home to raise a family now reentering the workforce 10, 15, 20 years after they left it.

I raised a question earlier about the pipeline. If the theory is correct, then where are the women? A glass ceiling of sorts still exists, and while the supports may not be made of concrete any more, the hurdles to get to the top are still, in many cases, substantial.

As I mentioned at the beginning, here are some key questions to ask of your leadership and of your organizations, to determine what hurdles remain. What is the atmosphere like in your organization? Is this a place where women come to work and stay? What's the retention rate? What do focus groups of employees say about the culture at your shop? Do you have women in leadership roles scattered throughout your place of work? Do you have to scramble to find them in order to make the right roster for an important committee? Are you using the same ones over and over? Do you have people willing to be mentors and coaches at work? With whom are these people working today—people who look just like them? What about the policy manual—does it still have 1980s dust on the cover? With the basics of good hiring and reasonable pay practices in place, if you've answered most of these questions positively, you probably have a pretty good place that entices women to come to work. If not, perhaps you have something to consider when you return.

To be sure you have a workplace that will enable women—and, for that matter, all employees—to be successful, here are some critical conditions for success. There has to be a way to measure and reward success that teaches others what success looks and feels like in your particular organization. There should be mechanisms in place to eliminate subtle barriers. After 20 years of news, most places have gotten the message and taken down the blatant, obvious barriers; but in many instances, subtle ones have crept in to replace them. Are there support structures in place? Are people using them? What about career reward and recognition—how do you reward people for doing good work? Is there a clear career path for them to follow? Is your company or department a place where employees can achieve their version of a comfortable balance between work and life?

In summary, we've seen data that say the pipeline is full. The critical issue today is how to retain talented scientists in industry. Atmosphere and other more tangible things that make up an organization—such as policies, external forces, and demographics—are all key indicators for retention that should be taken into account when reviewing organizations for their success in retaining talented women.

Today, women can be found at all levels of industrial organizations. We still have issues with promoting and retention. But—without a magnifying glass and without having to struggle too much—you can find women in many places in industry today, although not in the numbers nor at the higher levels of organizations that one might expect. There are many opportunities for improvement, even so; industry certainly is not nirvana. What about capturing the entrepreneurial spirit *inside* the halls of corporate America instead of driving it outside? Can corporate America dismantle some of its trappings that cause talented women (and other talented employees) to depart? Will there be an imbalance in supply and demand of scientists and engineers in the future workforce?

I hope this overview has been food for thought and will prompt some energetic, healthy discussions at the breakout sessions later today. Thanks for your attention, and I'd like to open the floor for discussion.

DISCUSSION

Maureen Chan, Bell Laboratories (retired): Being one of the women with 30-plus years of experience who is now off the charts, I wanted to make a comment about that break in salaries, especially as it gets to such a gap at the 20-year range. I think that gap, at least in part, can be due in industry to the way in which straight raises were structured over the years. They are very cumulative.

Merit review may be done separately from salary review. Someone might get an excellent merit review and 120 percent of their allotment as a result of that review; but when the salary review came, because of a lower starting salary or perhaps a break in service due to illness, maternity leave, and so on, the raise allotment for that individual would be low. So a lower raise could go to someone more highly rated because that's the way the salary structure worked—it is (or was) cumulative, and I think the combination of that cumulative structure and lower starting salaries for the older women just has never been solved.

Shannon Davis: I agree. It is hard to make up that differential once you already start at a lower base.

Margaret V. Merritt, Wellesley College: One of the points I think you were making about the pipeline being full is that there is not a shortage of people for these particular positions. I would argue with that basic premise. I think that in his presentation this morning, Arthur Bienenstock was suggesting the pipeline is going to be empty, that in fact we are not replacing people within that pipeline. I would also conclude that 20 percent of women in the pipeline does not represent the proportion of women in the population. Consequently, I think that one really has to work on the pipeline issue, particularly given the information presented this morning. I don't know if you have any comments on that.

Shannon Davis: Part of my bias exists from having lived in the pipeline for the last 15 years and the problems that I have encountered (as opposed to the problems affecting the entry-level positions). You can pick a 20 percent level, you can pick a 10 percent level, you can argue that women are 50 percent of the population so we ought to be 50 percent of the degrees that are granted; I am a lot more interested at this point in the cracks in the pipeline and where those people are going, because no matter what number you pick you ought to see those numbers track through—wherever they are.

Margaret Merritt: I agree with that particular notion, but I would like to make sure that we are not forgetting the entry people.

Shannon Davis: Absolutely. We need to have them.

Janet G. Osteryoung, National Science Foundation: You talked about where women aren't. I would like to mention something that everybody is aware of but often people don't think about, and that is the environment—particularly in a large company or in a government agency—the environment in which people work. I have worked in a lot of different organizations, and I have never seen a white male cleaning the toilets. In fact, it is not only almost always a woman, but a woman whose skin color is not lighter than the paneling in this room, and the same is true as you go up through the organization. The lowest levels are grossly overpopulated with women and with minorities of both sexes, and I think it is very difficult for people to think constructively about problems like this when they work in that kind of an environment.

Maria Spinu, DuPont: I think I am just optimistic. I really believe that in the future you will see a more realistic picture of the women chemists and what they have accomplished in a 20- or 30-year period. I know a lot of senior colleagues with wives who have earned a chemistry degree but have never practiced chemistry because they chose to follow their husbands to different locations to fulfill their husbands' careers. I think these days we, the women, are much more career oriented as well. So, the 20- to 30-year picture is going to look different when we look at women chemists who started their careers in the 1980s and 1990s.

Shannon Davis: That would be nice.

Victoria Friedensen, National Academy of Engineering: That was a really good presentation and dealt very well with a lot of complex issues. To go back to the first comment, however, about the pipeline: I think 20 percent is a very fair sort of an ideal. However, I would like to point out that more and more young women in the United States, especially young white women, are taking AP courses in high school. Girls are graduating from high school at ever higher rates, and their SAT scores, especially their math scores, are climbing incrementally. We have more kids who could, indeed, become chemists and chemical engineers going on to college, and yet the enrollment rates for chemistry and chemical engineering are plateauing. So while we are doing a very good job of increasing the numbers of engineers who are women, we are losing the market share of women who could be engineers.

One final comment about things to do in industry. If management is not made accountable for the success of the organization's people, you will not see women rising through the ranks. I think one of the first bullets ought to be a written policy in which senior managers are made accountable for the retention of women and underrepresented minorities on the staff.

Nancy B. Jackson, Sandia National Laboratory: I work in the Defense program's laboratories, and if you think it is hard to get women into the chemical sciences, you should try nuclear weapons. We have an outstanding woman who just became our executive vice president, which is a real treat for all of us. One of the first projects she took on addressed the difficulty at Sandia of attracting and retaining women in the first line of management, and she began a study in this area. She sent a team out to look at other places—Motorola, Los Alamos, and a number of other places that have different sorts of records with women—and she held several internal workshops.

One unexpected and interesting finding was that, in general, the women at Sandia felt that at the first level of management the managers were so bad that the women didn't want to be a part of that group. The women didn't have the same ego needs or perhaps salary needs that the men did, and so they didn't feel it was worth it to become a manager. Sandia is learning that one way to go about trying to attract more women into the first line of management is to be a little stricter about what they do with and accept from their managers, both male and female.

Donna J. Dean, National Institutes of Health: A couple of weeks ago at an NIH-related activity focusing on underrepresented minorities in science and medicine I heard a very good phrase related to the glass ceiling concept: the sticky floor. Even in places where there are no glass ceilings, there is still a sticky floor holding people back.

You were talking about people at the entry level—let's call it the bachelor's-degree level—that industry pulls out of the pipeline into industrial positions in chemistry, the people that you are actively recruiting. If you don't hire those people at the entry level, and if your counterparts in industry don't hire them for a chemistry position, what do those bachelor's degree graduates in chemistry and chemical engineering do? Do you have any idea?

Shannon Davis: That is a really good question. I don't know where they go. If some of the numbers that we had are accurate, I am perplexed. I don't know where they go.

Sally Hunnicutt, Virginia Commonwealth University: They go into health-related fields, I think. Probably if you removed them from the pool of bachelor's degree recipients in both engineering and chemistry, you would find that the numbers of women going into chemistry are lower. I don't know what the numbers are, but isn't it closer to 50:50 now in medical schools?

Donna Dean: That's not really my question. I am presuming that the people who are going to graduate school or medical school are not the people who are in the front lines applying for industrial positions at the bachelor's level.

Shannon Davis: You're saying that bachelor's-degree engineers were intending to go into industry to start with and if we don't hire them, what do they do?

Donna Dean: Right.

Shannon Davis: I haven't a clue.

Victoria Friedensen: They go into marketing.

Shannon Davis: They go into marketing and sales. They get their MBA and join an Internet start-up firm.

Participant: There is a big divide between chemistry and chemical engineering.

Shannon Davis: It is huge.

Faith Morrison, Michigan Technological University: Actually, almost the same question had been rolling around in my head, but it's a little different because I think the answer to your question is that they are underemployed in some parallel field. I can name you five black women chemical engineers from our institution who were unable to find an entry-level job in the last 5 years. People think that everybody who gets a bachelor's of science degree and perseveres to the end in chemical engineering gets an entry-level job in their field of choice, but it's not true. So, you can't take the number of degrees awarded as the number of engineers produced. There is an enormous leak in your pipe at the very beginning: some of it is due to low performance on grades and then using GPAs as the cutoff for hiring and some is disaffection.

My sister-in-law, who is a chemical engineer, never even pursued a job in chemical engineering. By the time she finished, she was so disgusted with the field that she wiped her hands clean and just never even tried. I have anecdotally encountered other people who have for whatever reasons—climate or atmosphere—just decided, This is not for me. Those people drop right off your charts, so you can't count them anywhere along the line. They never pursued this career.

Catherine Woytowicz, American Chemical Society: One thing that I think is really important is, Why are we "taking it"? I haven't heard one person get up here and say, "I am going to do these five things when I get home."

I urge you to do a couple of things when you go home today. First, that five-page pink handout that you have is a list of biographies. Don't take it home and make five pink paper airplanes. Take it home and contact some of those people. If you need someone in industry, find an industrial contact from that list. If you need someone in academia, find somebody from that list. If you need someone in the government, find somebody from that list. Vow to yourself that you will not throw that packet somewhere and let it collect dust. Don't put it in a drawer. Don't put it in a pile on your desk. Take that handout and make it a priority to contact one person from that list and start a network. If you don't have a network where you are, start making one. Stop taking this kind of thing.

This goes back to mentoring, too. Find one person in your department or in your community that you can mentor. If you need a mentor, find one. I was going to ask, Where were you when I was in grad school? Where were you when I was an undergrad and I really needed you? But the truth of the matter is, I still need you now. I am early in my career, and I could use a mentor—a better mentor, maybe, than I've had in the past. I have had a lot of men who haven't taken an interest in my career, and now I have butted my head against some of these glass ceiling issues, partly because I haven't taken advantage of a mentor.

That's what I am going to do for myself, and then network, network, network. Get out there and do something about it. And, if you say that language makes a difference, make a commitment to use "he or she" and inclusive language yourself. And stand up for your colleagues. If one of the women in your department has a problem, make sure you stand behind her. We are all guilty at times of saying, "I had it hard. She should have it hard, too." It doesn't have to be that way.

If we want to encourage people to stay and we want to retain young women, especially talented young women, we had better get behind them and push. If it's true that the pipeline is absolutely fine, then let's get some caulk and patch the cracks.

Jong-On Hahm, National Research Council: It may be that the pipeline *is* a problem. Retention is a particularly acute problem for chemical engineering at least; I saw a statistic a couple of years ago indicating that chemical engineering had the lowest rate of participation in the engineering workforce by its women graduates. It was only half, and this was the lowest of all engineering fields. So, if 25 percent of the degrees were going to women, only 12 percent were working in the chemical engineering field, and again this was the lowest of all the engineering fields. If somebody could give me a good explanation, that would be wonderful.

E. Ann Nalley, Cameron University: I'll tell you where those women are: They have husbands who have jobs. They are not mobile. They don't go to the big chemical companies; they go to the small companies—there are lots of small chemical companies in Oklahoma, in Kansas, and in New Mexico. They are the chemists at the wastewater treatment plant. They don't get surveyed. And they are not members of the American Chemical Society.

I am on the ACS Board of Directors and I have tried to recruit them, but they don't see that they need to be members. They are out there, but they cannot move, and they cannot go to the big companies. They don't go on to graduate school, although most of the people who graduate with B.S. degrees in chemistry are practicing chemists. They are the Texas State chemists. They don't get paid very well, but they do have jobs in chemistry, and I wish somehow we could include them in our survey.

I agree, also, that we need the pipeline to be expanding. If we don't get the young girls into it, then our pipeline is shrinking, not expanding. I don't agree that the pipeline is full. It has a long way to go before it is full.

Shannon Davis: I have to agree with Ann. I would love to see those people included in the surveys, because the more research I did the more concerned I got. I think the pipeline ought to continually grow bigger and bigger; but I really also want to applaud the woman who said, "Let's go find some caulk." Let's go find some caulk, folks.

6

Title IX for Women in Academic Chemistry: Isn't a Millennium of Affirmative Action for White Men Sufficient?

Debra R. Rolison
Naval Research Laboratory

The subtitle of this talk—"*Isn't a Millennium of Affirmative Action for White Men Sufficient?*"—serves as an exercise in putting the shoe on the other foot. The first university, the University of Bologna, was established, according to its Web site, in 1088.[1] The actual date is debatable because one has to decide when the "modern" university—as a forum for open discussion and as a place to seek new knowledge—arose out of the ecclesiastical schools; but the University of Bologna pegs its establishment at 1088. During most of the ensuing millennium, universities blatantly practiced preferential hiring: men only.[2] It is only recently that the hiring preference has started to change.

One might ask if a millennium isn't enough time passed before we diversify our university system. In the United States, a country blessed with a highly educated, diverse populace and where the modern university, public and private, lives and breathes by federal taxpayer dollars, one can rightly ask, Should the American taxpayer still support institutions that continue to hire white men preferentially?

Dan Greenberg, who writes a column in the *Washington Post*, touched upon something very similar one year ago: "Universities claim to be sources of enlightenment and progress. Of course they are, but in important respects they are also retrograde institutions detached from a society that is increasingly resentful about their ways."[3] Greenberg had numerous "retrograde" examples, some of which were tied to the MIT report that was issued in 1999 describing discrimination against the senior women faculty in MIT's College of Science.[4] Greenberg faulted U.S. universities, in general, for the slow progress of women's academic careers, especially in the sciences.

---

[1]The history of the University of Bologna, including the presence of women scientists and legal scholars on its faculty during the Renaissance and the Age of Enlightenment, can be traced on the university's Web site, <http://www.unibo.it/avl/english/story/story.htm>.

[2]Unlike most European universities, the Italian universities, and especially the University of Bologna, did have women faculty before the 1800s, including the outstanding physicist Laura Bassi (see <http://www.unibo.it/avl/english/biogr/bio15.htm>), but the faculty remained predominantly male.

[3]Daniel S. Greenberg, "Where higher ed has it backward," *Washington Post*, 8 June 1999, p. A19.

[4]"A study on the status of women faculty in science at MIT," *MIT Faculty Newsletter*, March 1999, available at <http://web.mit.edu/fnl/women/women.html>. This article summarizes a 150-page unpublished report on the suject that was prepared in 1994.

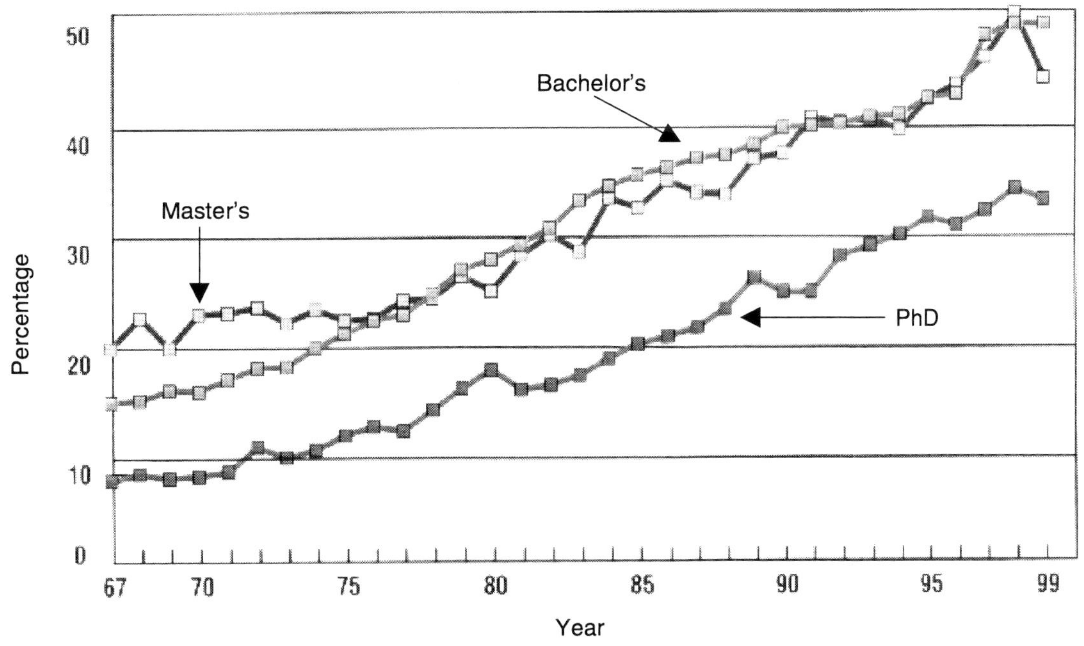

FIGURE 6.1 Percentage of chemistry degrees earned by women from 1967 to 1999. SOURCE: These figures are derived from the yearly starting salary surveys performed by the American Chemical Society. The figures for 1999 appeared in *Chemical & Engineering News,* March 13, 2000.

In a recent study of the paradox of the critical-mass issue,[5] the authors cited a suggestion from B. Lazarus, a representative of WISE (Women in Science and Engineering), who advocated that the National Science Foundation (NSF) cut off grants to those universities that don't have a minimum number of female faculty in their science and engineering departments. That is a pretty drastic suggestion, but one can make it more drastic—cut off not just NSF funding but all federal funding.[6]

THE QUESTION UNDER CONSIDERATION

Should federal funds—not just NSF-derived, not just NIH (National Institutes of Health)-derived funds, but all federal funds—be withheld from those universities that do not increase their faculty hires to reflect the diversity of today's pool of U.S.-granted chemistry Ph.D.s?[7] As seen in Figure 6.1, women composed one-third of that pool in 1999 and have been present at more than 20 percent since 1985. Despite that year-in, year-out production of talent, NSF reported in 1998 that only 12.5 percent of the senior faculty (associate and full professors) in the natural sciences and engineering at U.S. universities and 4-year colleges were women.[8]

---

[5]Henry Etzkowitz, Carol Kemelgor, Michael Neuschatz, Brian Uzzi, and Joseph Alonzo, "The paradox of critical mass for women in science," *Science,* 1994, *266,* 51.

[6]Debra R. Rolison, "A Title IX challenge," *Chemical & Engineering News,* 13 March 2000, *78*(11), p. 5.

[7]Debra R. Rolison, "A Title IX challenge," *Chemical & Engineering News,* 13 March 2000, *78*(11), p. 5.

[8]National Science Foundation. *Women, Minorities, and Persons with Disabilities in Science and Engineering: 1998.* NSF 99-338, Arlington, VA, 1999.

Is it time, in other words, to convince Congress (or the Judiciary) that the loss of federal funds can be a Title IX action against U.S. chemistry departments and their universities for their entrenched "inability" to hire women?

WHY TITLE IX?

The Education Amendments of 1972, commonly called Title IX, state in the first subsection (A: prohibition against discrimination): "No person in the United States shall on the basis of sex be excluded from participation in, be denied the benefits of, or be subjected to discrimination under any education program or activity receiving federal financial assistance."[9] I would argue being a professor in a federally funded university is an educational activity.

Subsection B, which discusses preferential or disparate treatment, includes language that states that Title IX cannot be used to discriminate. This subsection does permit, however, "the consideration in any hearing or proceeding under this chapter of statistical evidence tending to show that such an imbalance exists."[10] An imbalance exists, and has for over 15 years, between the fraction of women graduated with Ph.D.s in chemistry and the fraction of women who have applied and been hired for faculty openings in U.S. chemistry departments.

Now, Title IX is a very big hammer. I think a very plausible case can be made for using that hammer, but the laws usually applied to discriminatory work environments are those embodied in the Civil Rights Act and Equal Employment Opportunity legislation.[11] This legal recourse means that one files a lawsuit against a specific workplace, and if the treatment is especially egregious, the lawsuit can be expanded to a class-action suit. Why not fight this problem—the acknowledged lack of women in the academic chemical workforce—through lawsuits taking one chemistry department at a time? My response is that women should not be asked to wage this campaign lawsuit by lawsuit by lawsuit, because that is a war of attrition—against the women. The *system* is broken. The women are not.

Here is an example of the futility of one-chemistry-department-at-a-time lawsuits. One of the best known examples in chemistry is the Rajender case at the University of Minnesota. Dr. Shyamala Rajender was performing the duties expected of tenure-track faculty in the university's Department of Chemistry, but the university refused to consider her for a tenure-track position. She filed a lawsuit in 1973, alleging sex discrimination. That lawsuit was expanded in 1975 to the entire University of Minnesota system on behalf of all academic, nonstudent, female employees of the university. In 1980, the university signed a consent degree that agreed to policy changes, the scope of which is said by legal experts to be unmatched in the higher education system.[12]

---

[9]A list of exclusions, listed in subsection A of Title IX, includes certain exemptions for specific organizations, such as religious and vocational and (my favorite) beauty pageants. The Department of Labor's Web site, <http://www2.dol.gov/dol/oasam/public/regs/statutes/titlexi.htm>, describes Title IX in detail.

[10]See the Department of Labor's Web site, <http://www2.dol.gov/dol/oasam/public/regs/statutes/titlexi.htm>.

[11]Information on the federal laws, through Title VII of the Civil Rights Act, that prohibit employment discrimination based on race, color, religion, sex, or national origin, can be obtained at <http://www.eeoc.gov/facts/qanda.html>, the Web site for the U.S. Equal Employment Opportunity Commission.

[12]Nijole Benokraitis and Joe R. Feagin, *Modern Sexism: Blatant, Subtle and Covert Discrimination,* 2nd ed. (Englewood Cliffs, NJ: Prentice-Hall, 1995), available at <http://www.runet.edu/~Iridener/courses/getrid.html>. The Rajender case is also summarized on the Web site for the Minnesota Women's Center, <http://www1.umn.edu/mnwomen/mwchistory.html>.

The decision against the University of Minnesota looks like a clear verdict that women were being discriminated against by the university. What was the outcome? A lot of money was spent: $7 million in 1980 dollars in legal fees and settlements, including $1.5 million to Rajender's lawyers. And the departmental faculty today? If you check the University of Minnesota's chemistry Web site[13] and count the number of assistant, associate, and full professors—excluding the professors emeriti and colisted faculty—46 professors are listed, 3 of whom are women. The University of Minnesota lost that lawsuit badly, and 20 years later women represent only 6.5 percent of the faculty in the university's Department of Chemistry.

This is why I believe that fighting this fight one department at a time is brutality against the women involved.

FINALLY! ROOM IN THE ACADEMIC POOL—A HISTORIC OPPORTUNITY

Employment opportunities for Ph.D. scientists have only recently become healthy again, so until now women who wanted to go into academia had a very tight job market in which to try to make their way into that workforce. Today, all the people hired in the 1960s, during the boom years for expanding the scientific enterprise in this country, are nearing retirement or emeriti status. For the first time, there is room in the academic pool.

This egress and accompanying ingress of faculty offers the chemical sciences a historic opportunity. Faculty positions are opening up and will continue to do so for the next 5 to 10 years. Are we, as a profession, going to seize this opportunity or are we going to squander it? Will a woman-free (or woman-sparse) faculty be locked in for another generation? If the number of women on chemistry faculties is not substantially increased during this historic opportunity, every chemist in this country should be ashamed.

WHY ARE THE NUMBERS OF WOMEN ON CHEMISTRY FACULTIES SO LOW?

One clue emerges in the contrast between cocktail folklore and real statistics. When one talks to (primarily male) chemistry faculty at professional gatherings, one is usually told that 10 percent of the applications for a new faculty opening came from women. This number is pathetically low. Why is the applicant pool not 20 percent, 30 percent, or $33^1/_3$ percent female, when for every two men to whom we give a Ph.D. in chemistry in this country, there is one woman? Why does the applicant pool so poorly reflect the actual candidate pool that exists?

As Figure 6.1 reveals, since 1985, more than 20 percent of the Ph.D.s in chemistry in the United States went to women. That span of 15 years covers more than two tenure cycles. The women are there, and they have been there for years. Why are they not even *applying* for faculty positions? The language I like to use is that they are voting with their feet—voting against the university.

Graduate students working toward a Ph.D. constitute the pool of candidates that ultimately chooses for or against the university as a career home. These doctoral students have spent 4 to 6 years working hard, trying to master the craft of chemistry and trying to hone their chemical creativity. They have spent far less time or thought on how to create a career in chemistry. The one institution known with any certainty to them is academia. They know that chemists get jobs in industry, but do they really know how to make a career in industry? They know next to nothing about the government labs.

[13]The Web site for the Department of Chemistry at the University of Minnesota can be found at <http://www.chem.umn.edu/ directory>.

The graduate students of today also do not know much about career possibilities in small entrepreneurial companies and dot-coms, but—then and now—graduate students do know academia. More and more, men and women—and the women more than the men—are saying, "No, thank you" to a life in academia. Why? To answer that, I have borrowed a list of reasons from the speaking package assembled by the Committee on the Advancement of Women Chemists, or COACh.[14] Part of this list is a realistic look at why women might not want to go into academia; some of it is tongue-in-cheek, and by being tongue-in-cheek it is even more spot on. Are women choosing not to go into academia—are they voting against a university career—because:

- The hours are long and the pay is low?
- They don't want to serve on so many fascinating university committees, including every affirmative action committee ever constituted on campus?
- They don't wish to compete for grants when receiving four Excellents and one Very Good means their proposal is dead?
- They don't want to postpone having children until they have gained tenure? (Or, as one of my colleagues at the Naval Research Laboratory likes to say, until they are almost postmenopausal?)
- They don't want to be role models for every young woman on campus and beyond? All too often being a role model means being a mentor and adviser not only to a disproportionate number of students—male and female—but also to male colleagues seeking a friendly and unthreatening ear.

Or is it, as one male chemist commented, because once again women have shown that they are smarter than men? Some things in this list are funny but true, and some are sad but true.

I want to expand on this aspect of role model and mentor. That role is an enormous burden on one's energy, and it is a burden that many women take on gladly; but one person really cannot mentor the planet. After I proposed, in an editorial in *Chemical & Engineering News,* that it was time to apply Title IX to chemistry departments and presented a truncated list of reasons why that was a rational proposal,[15] I received reams of supportive e-mail from both men and women—students and faculty, as well as people in industry and government.

One of my electronic correspondents, Anna Farrenkopf, who is a postdoctoral researcher at the Oregon Graduate Institute, is also a leader in a mentoring program at OGI called Advocates for Women in Science, Engineering, and Mathematics (AWSEM). The AWSEM program brings young students to OGI over the course of the year, not just as a one-time event, to see how women do science: Anna and her compatriots have the students do experiments and show them what an exciting career science and engineering can provide.[16] Some of Anna's science can be seen in Figure 6.2, in which the tank is actually a huge electrochemical cell containing a slice of the Columbia River. In this test bed, Anna has planted a microelectrode so that she can study the electrochemical properties of the sediment using a probe that has a characteristic dimension of about one micrometer.

[14]The Committee on the Advancement of Women Chemists (COACh) exists to increase the number and effectiveness of women in research and leadership positions in the chemical sciences. Seed funding to establish the committee was provided by the Camille & Henry Dreyfus Foundation. For more information on COACh and implementation of its strategies to enhance the leadership effectiveness of women chemists (including development of job skills for women already established as associate and full professors), contact co-chairs Geraldine Richmond, of the University of Oregon, <richmond@oregon.uoregon.edu>, and Jeanne Pemberton, of the University of Arizona, <pembertn@u.arizona.edu>.

[15]Debra R. Rolison, "A Title IX challenge," *C&EN*, 13 March 2000, *78*(11), p. 5.

[16]When women discuss their science, it makes the point far better than almost anything else that women do great science. Our energy, quality, and devotion are an important message we transmit when, as women in science, we talk about our work.

FIGURE 6.2 The rotating annular flume surrounded by Advocates for Women in Science, Engineering, and Mathematics students. Microelectrodes are used to track chemical changes (fate and transport) across the sediment-water interface subsequent to resuspension of the sediment; the microelectrodes are obscured in the flume by the suspended solids in the river water overlying the Columbia River sediment bed. Photo taken by middle-school science teacher Laurie Denio.

Anna, the postdoc, read my "Title IX" editorial and wrote: "Hi, Debra. You go, woman. Wow. That was quite an editorial. I made 100 copies and put it in all the faculty and postdoc boxes here at the Graduate Institute." I asked her to give me some feedback since she had taken that initiative, and she wrote back, "Well, one of my male colleagues walked in with a copy of the editorial and said, 'Anna, this is for you.' I looked at him and said, 'No, *I* put it in *your* box.' "

Dr. Farrenkopf then expanded on that disconnect, about what it means to be a woman surrounded by people who need such reminders from women. "Everywhere I go, I am told that working as a woman in science is my issue. It isn't one to be shared by all of my colleagues. It is mine alone." Anna's observation gets back to the point that women continually have to make to men: Women alone aren't supposed to solve this problem. Women *and* men have to solve this problem.

Anna continued: "It is funny: the same folks who dismiss the women-in-science issue as pertaining exclusively to me are the folks who send me their graduate students, postdocs, and staff to mediate or consult with me when perceived or imagined issues of gender or gender equity get raised. I definitely know that I am not alone, for quite often my workday consists of supporting a network of contemporaries and younger men and women." This level of mentorship is being asked of a postdoctoral researcher.

WHY ARE WOMEN VOTING WITH THEIR FEET AGAINST ACADEMIA?

In my view, women are the canaries in the mines.[17] The fact that women are not applying for academic openings in proportion to their presence in the candidate pool tells you that our chemistry departments are not providing an environment in which a human being can grow and thrive.[18]

What if we take the relative absence of women in the faculty applicant pool as a tacit statement from women that chemistry departments are not healthy places for human beings? Universities and chemistry departments are, as Dan Greenberg said, "retrograde"[19]—they are missing representation from women and minorities and do not serve a modern society.

The current departmental environment is unhealthy in several ways:

- Unhealthy for men and women who want children—and who want to play a meaningful role in their lives, not just give birth to or sire them;
- Unhealthy to the women who, once they demonstrate productivity, scholarship, and mentorship, still reap less respect than their comparably productive colleagues and less of all those things that indicate that one has respect: awards, raises, extra space, the right kind of committees;[20]
- Unhealthy to those men and women who want to create teams, who want to do collaborative, cooperative research rather than cutthroat research;
- Unhealthy to those men and women who think that the reason they are in a university is to educate and mentor and guide students; and
- Unhealthy to those undergraduate students who are trying to picture what their life will be like as a chemist, especially as a chemist in academia, when what they often see are people they have no desire to be like.

This list makes a perfectly sound case that chemistry departments, as we have configured them, are not healthy environments for human beings, and that is why the women are ignoring these departments as a career home.

If the absence of women from the applicant pool is the observable, what is the mechanism? I would postulate that the problem lies with departmental and scientific culture, which in turn exists within this huge bubble of our societal culture. Wolf-laureate Chien-Shiung Wu, a physicist who arguably should have gotten a Nobel Prize but didn't, addressed that point before she died: "I sincerely doubt that any open-minded person really believes in the faulty notion that women have no intellectual capacity for science and technology.[21] Nor do I believe that social and economic factors are the actual obstacles that prevent women's participation in scientific and technical fields. The main stumbling block in the way of any progress is and always has been unimpeachable tradition."[22] This observation came from a

---

[17]Debra R. Rolison, "A Title IX challenge," *C&EN*, 13 March 2000, *78*(11), p. 5.

[18]Joan Selverstone Valentine captured the essence of this problem (*Science* 1992, *256*, p. 1615) by telling young women: "Shun departments with a record of hostility to women (and to assistant professors in general), no matter how high those departments may stand in national rankings."

[19]Daniel S. Greenberg, "Where higher ed has it backward," *Washington Post*, 8 June 1999, p. A19.

[20]This manner of disrespect hearkens back to the MIT report (see footnote 4). Students and young research scientists are not unaware of the treatment of senior women staff and faculty. Until a few years ago, my postdoctoral associates at the Naval Research Laboratory said that there was no way they would consider an opening at NRL because of how I was treated relative to my male colleagues.

[21]Unfortunately, many women chemists could easily generate a list of men chemists who feel that women have no intellectual capacity for science, but we will give Professor Wu that part of her statement.

woman who left Shanghai and Chinese culture for the West, so she knew something about unimpeachable tradition.

What has our tradition been? In Western science it has been, since day one, a world without women. David Noble wrote a book about that.[23] In fact, his book jacket shows Albrecht Dürer's very famous etching of Adam and Eve—*without Eve*. Noble had this glorious piece of Dürer's art retouched by Kathy Grove to take Eve out of the picture in order to symbolize, all too accurately, Western science.

Western science derived from the intellectual context of monasteries and ecclesiastical schools and then moved into the universities, retaining almost all the vestiges of its beginnings in the monastery.[24] The ideal was, and in some ways still is, that one is dedicated around-the-clock to the scholastic pursuit of knowledge, and to do that either one must be a monk (or exhibit monastic dedication) or have an infrastructure. For most of our recorded civilized history, that infrastructure has been a wife, or as Garvan Medalist Janet Osteryoung has mused, "Every professional needs a wife."

It certainly is not an option most women have open to them in today's world. What the university requires is disconnected from what modern society has become—a world with women and minorities. As members of a profession, we should step back and ask if we want one of the most rewarding paths of our profession to be off-limits to men and women because they want their personal, human priorities to be given a place, too.

That disconnect—between what was and what is—is a major problem. But the current environment in departments is a multivariate problem—improving the environment will require more than one solution, even if Title IX is probably the biggest hammer we can take to it.

COVERT RATHER THAN OVERT BIAS

Recently, Howard Georgi, Mallinckrodt Professor of Physics at Harvard, enumerated a number of things that many people have said before,[25] but now a white male physicist is saying it: "Unconscious discrimination arises due to deep-seated habits that will be very hard to change,"[26]—much of which gets back to the in-built gender and racial schema of our species.[27] Georgi continues: "Our selection procedures tend to select not only for talents that are directly relevant to success in science"—and one would hope that that includes brightness, creativity, the ability to persevere, and treating well those in our charge—"but also for assertiveness and single-mindedness. . . qualities that are at best very indirectly related to being a good scientist and that clash with cultural pressures. . . . It is not impossible to succeed as a scientist without being assertive and single-minded, but the system encourages and rewards people with these traits in a number of ways."[28]

[22]Sharon Bertsch McGrayne, *Nobel Prize Women in Science*, 2nd ed. (Secaucus, NJ: Citadel Press, 1998), p. 279.

[23]David F. Noble, *A World Without Women—The Christian Clerical Culture of Western Science* (New York: Knopf, 1992).

[24]David F. Noble, *A World Without Women—The Christian Clerical Culture of Western Science* (New York: Knopf, 1992).

[25]For example, Elizabeth Zubritsky, "Women in analytical chemistry speak," *Anal. Chem.* 2000, 72, 272A.

[26]Howard Georgi, "Is there unconscious discrimination against women in science?" *APS News,* 2000, 9(1), The Back Page, available at <http://www.aps.org/aspnews/0100/010016.html>; and *Who Will Do the Science in the Future?* (Washington, D.C.: National Academy Press, 2000).

[27]Virginia Valian, "Running in place—After thirty years on the fast track, women are still hobbled by the cumulative effects of sexual stereotyping—a bias that begins in infancy and persists even among the most enlightened employers," *Sciences-New York* 1998, 38 (1), p. 18; and Gerhard Sonnert, "Women in science and engineering: Advances, challenges, and solutions," *Ann. NY Acad. Sci.* 1999, 869, p. 34.

[28]Howard Georgi, "Is there unconscious discrimination against women in science?" *APS News,* 2000, 9(1), The Back Page, available at <http://www.aps.org/aspnews/0100/010016.html>; and *Who Will Do the Science in the Future?* (Washington, D.C.: National Academy Press, 2000).

What They Say Versus What They Do

Again, the crux of the problem appears to be the departmental culture as exemplified by its reward system. Professors are human beings—they are mammals who want to feel good and loved and respected and get fed well and patted on the head, so they respond to the operative reward structure. In my opinion, the reward structure of the university has evolved an organism—the chemistry department of today—that is extremely hostile, unhealthy, and brutal to people who want to do something besides chemistry around the clock.

What does the departmental reward structure say it rewards? The first duty of chemistry faculty is to the young people whom they have chosen to teach, mentor, and guide in the joys and rigors of chemistry. Their second duty is to quality scholarship. In my worldview, if you want to do chemical research and not teach, not mentor, not guide students in the joys and rigors of chemistry, you have no business being in a university. Go do research somewhere else. Students are not cannon fodder.

Whom do chemistry departments *really* reward? Those faculty who bring in the most overhead-bearing monies, who excel at promoting their science, and who are single-minded and aggressive. Many people—men and women—are not comfortable with these attributes being the most rewarded and respected professional expressions of competence.

A valid point can be made that the U.S. university system does in fact serve society very well, in many ways, and produces people who do great science. So what! Does this mean that the university won't serve society and science *better* when it changes and integrally includes women and minorities? Science will look and be different when we bring in people with different perspectives, abilities, and even different analogies through which to think about their science. As scientists, we are always thinking in analogy, but if we always dip into the same pool, we limit the richness of our analogies.[29] And, should U.S. taxpayers have to support a discriminatory institution? I think these are all fair questions.

REDEFINE WHAT DEPARTMENTS WANT

How can matters improve when 10 percent or less of the applications for faculty openings in chemistry come from women, although the candidate pool is at 33 percent? One way is to stop accepting the male dictionary as the operative lexicon. Women and men must challenge the standard terms and patterns and expectations. For example, most university faculty search committees are not really *search* committees; they are manila-envelope-opening committees. These committees do not seek out new life forms (i.e., women and minorities). If a life form sends in a manila envelope, the committee will open it, but the members of the committee are not going out there to search for new life forms. Universities understand that to build competitive functional teams, recruitment is absolutely vital—they would fire their basketball coach if he didn't do it. Chemistry departments also have to *recruit* what they need, and they need women. Chemistry departments already recruit the men they want as faculty.

Now the members of the manila-envelope-opening committees are the same people who have been generating the 20 percent—and higher—fraction of women Ph.D.s since 1985. Why have these faculty not wondered why more of their women doctorates were not applying to academia? That discrepancy should have made the faculty on the manila-envelope-opening committees wonder what that said about their department as a place to create a career. What does that lack of curiosity say about their ability as

[29]Londa Schiebinger, *Has Feminism Changed Science?* (Cambridge, MA: Harvard University Press, 1999).

scientists to observe and make deductions with respect to a natural phenomenon occurring right under their noses?

So our faculty search committees must *recruit* women and minorities and not just open envelopes. But if search committees really become *search* committees, one of the things they then need to recognize is that there is always bias in evaluating applicants' packages. Recruiting novel life forms is not going to help if they are evaluated in a way that makes it irrelevant that they applied.

GUARDING AGAINST THE MYTH OF OBJECTIVITY

Men need to get over this fantasy they have that they are objective. They aren't. As psychologists have shown,[30] women are not objective either, but more of us do seem to be aware that we are not objective, and we step back and admit, Maybe I had better give this (situation, person, application) another look. That recognition is an important step. We now have to start training men to recognize that they are not objective and that they, too, need to step back and look at the situation in slightly different ways. As Georgi points out, expand the evaluation criteria. Do not use the standard-operating-procedure list of criteria as the only way to rank candidates.[31]

Men also need to recognize a chemical truism: like dissolves like. Like likes like. Like does not like unlike. When something in an applicant's package reminds a committee member of himself, he immediately becomes a champion for that package—and most men are not going to see something in a woman's application package that reminds them of themselves. Let us look at some examples where women are deemed "unlike" and less qualified for the job.

This first example, reported in the *Washington Post* in 1997,[32] described the enormous increase in women appointed to orchestras once orchestras started holding auditions behind screens. Because the people evaluating the performances had no idea if a man or a woman was playing.

The second example comes from psychology, a field that is very aware that a gender schema exists and that men are overvalued and women are undervalued. Even psychologists are seriously hurt by gender schemas: a study showed that university psychology professors prefer two-to-one to hire Brian Miller over Karen Miller, even when the application packages under review are identical. Merely attaching a male name to the curriculum vitae raises the *apparent* quality of that application in the eyes of both male and female reviewers.[33]

In the third example, a study of fellowships awarded by the Swedish Medical Research Council showed that the productivity of a woman candidate applying for the prestigious SMRC fellowship had to be 2.5 times more stellar to get the same competency score as the average male candidate.[34]

It is clear that serious bias exists in how women are perceived in terms of their productivity and competence: women are consistently undervalued for their qualities and accomplishments. Search committees, department heads, deans, provosts, and university presidents need to be aware of this

---

[30]Virginia Valian, *Why So Slow? The Advancement of Women* (Cambridge, MA: MIT Press, 1998).

[31]Howard Georgi, "Is there unconscious discrimination against women in science?" *APS News,* 2000, 9(1), The Back Page, <http://www.aps.org/aspnews/0100/010016.html>; and *Who Will Do the Science in the Future?* (Washington, D.C.: National Academy Press, 2000).

[32]"Do unseen musicians get fairer hearings?" *Washington Post,* 13 July 1997, based on a report by Cecilia Rouse and Claudia Goldin to the National Bureau of Economic Research.

[33]Rhea E. Steinpreis, K.A. Anders, and D. Ritzke, "The effect of perceptions of gender in the review of the curricula vitae of job applicants and tenure candidates: A national empirical study," *Sex Roles: J. Res.* 1999, 41, p. 409 (reported in the *Washington Post*, 2 April 2000).

[34]Christine Wennerås and Agnes Wold, "Nepotism and sexism in peer review," *Nature* 1997, 387, p. 341.

cultural, unconscious bias and devise mechanisms to evaluate candidates (and faculty) and their productivity more fairly.

Search committees certainly know how to be proactive because they are when they recruit men: they check in with their colleagues and ask, Who is in your group, who's going to be the star? They must ask this of groups that graduate excellent women candidates. If only those professional colleagues are contacted who graduate excellent male Ph.D.s but very few women Ph.D.s, the search committee is not doing its job.

WIN THE BATTLE, LOSE THE WAR?

If search committees are converted into *search* committees, will that be wasted effort? If the problems enumerated above with respect to the unhealthy environment in chemistry departments are not solved, women will still vote with their feet and not apply to a university. To highlight the difficulty posed by the current environment of chemistry departments, I will now anonymously quote some of my electronic correspondents.

This first message was sent by a male professor with 19 years' experience: "A recent graduate student, the student who gave the best preliminary exam I have ever sat on—she should have been quizzing the faculty, absolutely fabulous—is headed for industry despite intense urging by me and her adviser and several others, proactively lobbying for this woman to go into academics. She doesn't need this mess. You cannot hire someone who doesn't apply." This stellar young scientist is not applying because she has treaded the water and she does not want to be in that particular pool.

Here is a discouraging observation from another male correspondent: "I do not sense that our best female graduate students are necessarily interested in academic careers. The strong economy and relatively low academic salaries may be partly to blame for this, but I think there is more." He hasn't quite yet acknowledged that the departmental environment is the "more."

I received the following message from a young woman who recently graduated from one of the premier groups at one of the premier California universities, who has just taken an academic job and chosen to dive into the pool. "This fall I conducted a job search for academic positions in chemistry. Many of the departments that had the lowest percentages of women were those that did not interview any, although there were certainly highly qualified candidates applying. I had a number of female colleagues with impressive academic achievements who applied this year. I think the departments that are most deficient in women are the ones most intent on staying that way, a difficult problem to remedy. I completely agree that without outside pressure it is likely that the situation will not change on a satisfactory timescale."

A satisfactory timescale is *now*, as we replace those scientists hired in the 1960s. Once that exchange of new-for-old faculty is complete, the door to academia will again be closed for a (long) while.

Here are some thoughts from another male correspondent: "I am on sabbatical with a small company and I think this is truly the future. This company is at least 50 percent women, and the upper management is also about 50 percent women. There isn't the slightest hint of friction or discrimination. I think this is where a lot of women are going and feeling very comfortable with their situation. It is extremely difficult for me"—and he is actively trying—"to encourage anyone into academics anymore, women or men."

And, of course, we have President Vest's realization: "I have always believed that contemporary gender discrimination within universities is part reality and part perception. True, but now I understand that reality is by far the greater part of the balance."[35]

The environment is the problem, and the environment is populated by men. As a completely different male correspondent said (and, yes, his tongue was in his cheek): "Saw the editorial. I was shocked—shocked—after all these years to find out that men were the problem. I never would have guessed."

It's not news. The men are the ones who run the system. They are the ones who are rewarded by the system. They are the ones who are going to have to agree to change the system. They will be helped by plenty of women, but they have to be a big piece of the efforts to change the system.

THREE OPTIONS FOR FIXING BROKEN INSTITUTIONS

First option: Complete demolition—start over.

Second option: Coercion—use a hammer or a very big stick. A very big stick would be the withdrawal of federal dollars. If the case can be made that chemistry departments have merited the application of Title IX, that will put universities in serious trouble, no doubt about it. And making a plausible case for Title IX appears possible. If so, where does the fault lie? Not with the women. They have done what was asked of them: stay in the pipeline long enough to get a Ph.D. Pumping more women Ph.D.s into the profession was long thought to be the start to the solution. But increasing the fraction of women with Ph.D.s in chemistry from 20 percent in 1985 to 33 percent in 1999 has not solved the women-in-science problem.

Third option: Change the reward structure—admit, as a profession and a society, that this system is still very vital, but that it must be changed. How does one change a standing structure? Change the reward structure. Do that and people change their behavior. Most mammals are not very adept at doing something just because it is the good, noble, moral thing to do. We do it because we are rewarded.

CARROTS ARE FOR VEGETARIANS

Another correspondent comments: "Title IX. I like the premise. It's just like with kids. If there is no consequence to bad behavior then what impetus is there to change it?" Yes, Title IX is a big stick. A number of other correspondents have offered suggestions that could improve the situation and perhaps avoid the stick. As more than one person wrote to me in the past few months: "I generally prefer carrots to sticks." My rejoinder: We are dealing with carnivores; carrots are for vegetarians. The strategies that these correspondents propose, if started years ago, should have moved the environment toward one more amenable to women—but if these strategies are implemented *right now* and given the time they need to work, chemistry departments are in danger of locking in woman-sparse faculties for another generation.

The system is broken, but the system actually went out of its way to break itself. Departmental culture evolved into its current form because of what was rewarded. What the department chose to reward was success at winning overhead-bearing grants, arrogance, and skill at self-promotion. That is what has been rewarded and cherished, and that is what most women want no part of.

[35]"A study on the status of women faculty in science at MIT," *MIT Faculty Newsletter*, March 1999, *11*(4), available at <http://web.mit.edu/fnl/women/women.html>.

WHAT WOULD HELP UNIVERSITIES ATTRACT MORE WOMEN FACULTY?

On-site day care, for one thing. University presidents may say that there is not enough space or enough money to make on-site day care happen for the numbers of male and female faculty who want it, but if the university gets hit with Title IX, there will be no federal dollars and there will then be no (or a greatly limited) university. The priorities of the administration will suddenly change. Universities have already demonstrated how fast they move when their federal biomedical research funds are held in abeyance due to violations of regulations regarding human subjects.[36] The priorities of the administrations must change to solve this problem.

For another, mentor the junior (and senior!) faculty in a way that illuminates career choices and that shows where the pitfalls and minefields are en route to building a respected career.

We could scale back the demands on faculty. Many of my faculty correspondents point out that they do too many things in the current university environment. People have always worked hard in universities, but now they are working insanely hard. The faculty members deemed highly successful by the reward structure have become the equivalent of CEOs—which requires working 80+ hours a week.

We could also bite the bullet and return faculty to their primary function, the reason we have universities in the first place: to train and challenge students and to conduct scholarly research that contributes to the greater store of human knowledge.

Lastly, change the reward structure. I have posited Title IX as a course of action[37] because I think punitive measures have been earned and paid for in full, but changing the reward structure also has much to offer as a strategy. Really reward those faculty who are there for the right reasons: for the students and for scholarship. The university that does so will inevitably alter its departmental environment to one that women will choose and one in which men and women will enjoy being citizens.[38]

People in academics do it right—we all know faculty who do. Let us start rewarding them. And why should we do that? Because a brutal environment—and I would, again, postulate that the typical U.S. chemistry department is a brutal environment—drains the joy out of doing science. Let me give you a personal example. A former postdoctoral associate of mine commented recently that it was a high tribute to my Ph.D. adviser, Royce Murray, that every person she had met who had come through his group—and she has now met about 10—retained a love of science, a desire to get in there and tussle with science and scientific questions. Coming from one of the Ivy League schools, she has seen more than her share of the browbeaten, and they are not joyous scientists. This country should want joyous scientists.

CAN THE FUNDING AGENCIES PLAY A ROLE?

A number of my correspondents have suggested that it is time to start funding graduate students outside the competitive grant-proposal-writing system. Faculty spend much of their time writing too many proposals instead of mentoring students. Others have suggested that if the federal government wants to step in and provide half the start-up package for women, strong but unfairly undervalued candidates will immediately be raised higher in the ranked list.

---

[36] Sheila Kaplan and Shannon Brownlee, "Duke's hazards–Did medical experiments put patients needlessly at risk?" *U.S. News & World Report,* 24 May 1999 (the Web version of the story can be found at <http://www.usnews.com/usnews/issue/990524/nycu/trials.htm>).

[37] Debra R. Rolison, "A Title IX challenge," *C&EN,* 13 March 2000, 78(11), p. 5.

[38] I predict that if the reward structure is changed, the system will turn on a dime.

A rather provocative suggestion comes from someone who served as a rotator at NSF in the early 1990s: If a graduate student costs $33,000 a year,[39] just fund every chemistry graduate student in this country at $33,000 a year. The faculty can then do their job without fear that the students will starve. If people want more than that, including postdoctoral researchers, then write a proposal. This correspondent also looked at the support of graduate students through NSF-funded proposals (in the early 1990s) and contrasted that with the total number of third- and fourth-year graduate students in the country. The dollar amount spent via NSF grants on students was the same number as would be spent funding every third- and fourth-year chemistry graduate student in the country—and all without writing one proposal. Those two pots of money are the same, but in the current scheme of things faculty have to work nonstop to ensure the funds are kept flowing and the students supported.

What if block-funding of graduate students were made available? Would that allow the funding agencies to focus on reviewing and funding the less safe proposals, the ones exploring nonmainstream problems, the ones so difficult to get through the current funding system? One can also wonder what such an arrangement would mean for student choice and market freedom. A block-funded graduate student would be the "buyer" and the professor the "seller," so the research program and environment would need to be appealing in order for the student to join the group. Might a marked change in graduate-school dynamics result?

With respect to the many proposals that any active researcher must write these days, the reality is, We have met the enemy, and it is us.[40] Researchers are trying to do more than they did 30 or 40 years ago. We tackle multiple problems requiring cross-disciplinary skills (if not teams of people) with multiple people per problem and all the high-tech instrumentation that that implies—ergo, we require multiple funding lines and must satisfy multiple sponsors. Gone (for most of us) are the days of specializing in one system, one technique, and exploring it with a few graduate students for many years. The science is more interesting now, I think, but we have contributed to the systemic breakdown.

CRITICAL MASS

The need to achieve a critical mass of women in chemistry departments is often mentioned as the way to solve the problems women face in academic science. But does critical mass magically solve the problem? Does getting women onto the faculty of science and engineering departments at critical mass—deemed to be about 15 percent—solve the problem? Only partly, according to Etzkowitz and coauthors,[41] who wrote, "The fallacy of critical mass as a unilateral change strategy is that female faculty pursue strikingly different strategies. They differentiate themselves because they have different goals." Women, like men, are not monolithic; we differ. We do things in different ways from one another—and we don't all like each other. And this differentiation among female faculty sometimes produces isolation even when the numbers on an absolute basis have reached the purported critical mass.

Etzkowitz et al. make another important statement in their paper: "Informal networks are indispensable to professional development, career advancement, and scientific progress. Contiguity of helpful colleagues improves the conditions for scientific achievement." Contiguity is the key. It is not just a

[39]The figure of $33,000 arises from the following estimate: a generic stipend for a graduate student plus tuition plus overhead, costs about $30,000 a year. Add another (burdened) $3,000 so that the student can do research, whether with chemicals or computers or a laser.

[40]Apologies to Walt Kelly and Pogo.

[41]Henry Etzkowitz, Carol Kemelgor, Michael Neuschatz, Brian Uzzi, and Joseph Alonzo, "The paradox of critical mass for women in science," *Science,* 1994, *266*, 51.

critical mass that must be achieved, it is a *percolation threshold.* For example, in a three-dimensional system in which the volume is filled with black balls and white balls, in order to achieve continuity of transport properties among the minority component, you have to reach a percolation threshold (a volume fraction) of either 15 percent or 18 percent.[42] Until the percolation threshold is reached, the minority component is isolated from the community of the whole. It is the *communication* between the minority balls that is critical to impart their character to the whole.

If a nominal critical mass is reached and 15 to 18percent of the faculty in a chemistry department is (finally?!) female but that critical mass does not communicate, it isn't a critical mass, because a percolation threshold has not been achieved. It is the communication that is important. In my opinion, critical mass as a singular goal will not solve the problem women face in science any more than keeping women in the pipeline has solved the problem. We are keeping women in the pipeline: 33 percent of our Ph.D.s in chemistry are now women, which helps—it *has* to help—but which has not solved the lack of women in academic chemical sciences. My concern is that if reaching a critical mass of women in an academic department is defined as the goal—the only goal—then achieving it may still not solve the problem of fully integrating women into the enterprise of science.

And a final concern: If we cannot pull women chemists into academics when we are graduating one woman for every two men, how are we ever going to get minorities on chemistry faculties?

ACKNOWLEDGMENTS

I would like to thank Janet Osteryoung and the Organizing Committee of the Chemical Sciences Roundtable for allowing me to expand on the 650-word editorial, written for the March 13, 2000, issue of *Chemical and Engineering News*, in which I proposed applying Title IX to U.S. universities for their "entrenched inability to hire women" in their departments of chemistry. I am deeply grateful for (and to) the numerous correspondents—students, postdoctoral associates, governmental and industrial scientists, professors, and nonscientist taxpayers—who sent and continue to send me their observations and suggestions with respect to the absence of women in the academic chemical sciences. Many of their comments have been included in this article. My greatest debt is to George M. Whitesides of Harvard University, for catalyzing my thinking on this topic and for providing a rigorous contrapuntal line: his half of the conversation considerably sharpened the shape and flow of my arguments.

DISCUSSION

Marylee Southard, University of Kansas: I have two questions. First, is there any way that we can get copies of your slides?

Debra Rolison: Yes, if your server will take a 6 MHz file you can get a copy from me.

---

[42]Richard Zallen, *The Physics of Amorphous Materials* (New York: Wiley-Interscience, 1998), chap. 4. The percolation model distinguishes between bond percolation, in which a pairwise connection exists between sites with bonds either connected or unconnected (percolation threshold is reached at 18 percent), and site percolation, in which all sites are connected and the sites, rather than the bonds between sites, establish the character of connectivity (percolation threshold is reached at 15 percent). If the analogy holds, and the number of connected bonds is the critical component for women in an academic department, then the "critical mass" threshold is reached at 18 percent.

Marylee Southard: Thank you. My second question concerns a system that is more completely broken, I believe, than what you had time to talk about today. This example is not in a chemistry department or a chemical engineering department but in a biology department that I know about. Six people were up for tenure. Two of them made it; the other four were two couples who didn't make tenure primarily because they didn't get a federally funded grant during their first 5 years. One of them was up for the highest teaching award conferred by the student body. Do you have any comments about this?

Debra Rolison: I think we often try to reassure ourselves that things are better for women in biology because the numbers are so much bigger. But two days after my editorial came out I heard from a plant physiologist: "Please come talk to us because we have the same problem." Clearly, critical mass has not solved the problem, and I agree it is broader than just chemistry and chemical engineering departments. I think we have to retrain universities as a whole to understand what their function is.

Linda McGown, Duke University: I loved your talk—you raised a lot of really good points. I would like to address one of these, the myth of the 150-hour work week, which I think is one of the things that keeps people from trying to participate in academics. In my own group I try to encourage my students to integrate their lives with their graduate studies. I think it is the best preparation you can give them for life after graduate school regardless of what area they go into. If they want to get married, I don't feel that they should have to ask my permission in the first place, but I go further and give them my blessing. If they (or their wives) have children, that is okay. I don't feel that I should have input in those sorts of decisions as long as they do their job.

Then the question is, What is their job? Is it face time or is it getting the work done? I have never seen that there is a significant difference between really good people who work for 40 hours and other people who are around for 90 hours. I just don't think that you can sustain that level of intellectual activity. I am lucky if I have 2 hours of serious original creative thought a week. The true gems do not occur often. You cannot force them. The rest of it is fluff, and I can do that fluff at home, at work, or in the car.

There are other kinds of things that you have to do, granted, although I don't think that is science, and I don't think that 90 hours makes a better scientist, although people have that perception. They believe that there is a linear track for certain times of your life. It is the medical school model as well, where you put off doing something—such as getting married—until you get your Ph.D. Then you put off kids until you are in your forties or fifties, and that doesn't work either. I had both of my children while I was in my tenure-track years. You could look at my productivity and the productivity of my colleagues, male or female, and I really don't think you could tell from the statistics if—or when—I (or they) had children.

I don't think there is a significant difference between my colleagues and myself. I think it's more an attitude of hazing in some respects and a culture that tries to make it as difficult as possible for people to do anything else— particularly if it involves time.

This is academia. We have the most flexibility of any job, and yet here we are saying that it's easier in industry to have children. I do not understand this. When my children were born, I was at home, and I had group meetings at home—we have computers now, and we have all kinds of means of communication—so I was gone for just a month or two. Now, I have never taken a sabbatical—which is my personal choice—but I have colleagues who were on sabbatical, and I could not tell because I hadn't seen them in 2 or 3 years anyway. These are all side tracks off the main issue of integrating your life into your career. It is so terribly discouraging that for some reason academia is not perceived as a wonderful place where you can think and ponder—do wonderful science and have a life. Instead it has been turned into a dungeon-like atmosphere.

Debra Rolison: Hamster on a wheel!

Linda McGown: It is really terrible, but I do think a lot of the problem is perception, because it hasn't been that way for me. I am not always happy, but I still choose to remain in academia. I have never opted out because deep down I love it, and I hope that I am an example to my students and to other people that you don't have to do things according to some schedule that other people give you. Just take the flak and tell your students to take the flak because academia is worth it. Others have no business telling you what to do with your time, if you are pulling your load.

Mary L. Mandich, Bell Labs: I really enjoyed your talk, and I think your proposal for withholding the funding would get right to the heart of this problem. At a recent lunch of professors in physics and chemistry, more than one professor suggested that distance learning is threatening higher education, especially really good higher education. I asked them why they thought that. They said, "Well, you know the quality of the teaching is really going to go down." And I said, "With all due respect to you folks in academia in this room, I've had some really lousy teaching at universities, and I've had some very good teaching. I would think you would get the same thing over the Internet—some really good course material and some really lousy course material." They said, "No, no, no, it's going to be really lousy." "Well," I said, "I have actually seen some really good stuff, so what we need is some way to sort through and find the really good stuff. Maybe the ACS and the APS could promote the good Web sites and help people find good material." To which a very distinguished professor said, "It would ruin universities because we would no longer have students to teach in chemistry and physics. They would go on the Internet, and our departments would shrink, and we would not get federal grants, the overhead of which is absolutely vital to the university." This is why I think going after funding goes right to the heart of where the university will be hurt and will have to change in order to survive.

Patricia L. Watson, DuPont: I have a comment about effecting change that comes from our experience at DuPont in the late 1980s and the early 1990s, when we were concerned about many issues in the workplace. Now, there are lots of professional issues, but as far as work and family issues were concerned, people at DuPont were working very hard. All the stresses that are seen in academia are certainly there in industry as well. We had a person, Faith Wool, in our human resources office who was an extremely innovative and assertive woman—in fact she was recruited by the Clinton Administration to help frame the Family and Medical Leave Act. Faith took the initiative to do several surveys at DuPont—very large and inclusive surveys—and after the second or third survey in which she had gathered a lot of information about retention and satisfaction, a lot of things started to happen around family and work issues.

What actually got people's attention—after they had absorbed the percentage of women that had considered leaving DuPont because of these issues—was the number of *men* who had considered leaving DuPont. It was at least half as large as the number of women. Until that point women were considered more of an issue because more of them had taken action to go to small companies and so on, as we have heard. But what I think precipitated changes at DuPont was when men in upper management noticed the number of men leaving. And I don't believe that at universities all the men are satisfied with the situation either.

Debra Rolison: Yes, we are hearing more about how unhappy they are with the situation.

Patricia Watson: Right. So getting some data on that could be very useful.

Debra Rolison: I would also comment that this is what made Ruth Bader Ginsburg's career: She sued to get family privileges for men. Women were having no luck in the courts, but she started to make the same points for men, and they succeeded. I think Joan Valentine said it best a long time ago: "When departments treat people well, women will do well in those departments."

Elizabeth Theil: I would like to comment on the stick approach as opposed to the carrot that was the NIH response on the issue of salary equity in the 1970s. I know, because I was a beneficiary under the Carter Administration, which encouraged salary equity. The NIH did surveys of PIs by gender at different institutions. They then wrote to the provosts of those institutions where the women were way out of line in terms of salary and threatened to withhold future NIH funding. That certainly got the attention of the provosts, and so the two women at my institution (North Carolina State University, where I was a faculty member for many years) who had NIH grants as PIs got salary increases for 2 years in a row. But the deans never bought into it, and as soon as the NIH went away, the salaries receded. So, it does work, but it has to be built in a way that provides continuity. I think the only way to do that is to have a money component that is also related to the requirement for critical mass. If you get a large enough number of women or men with the right ideas into a department with this threat, then when they are in place they will be able to make sure it continues. But if you just have isolated cases it won't work.

Debra Rolison: You almost need the entire chain involved. We all know departments that are really trying to bring women in—but the women aren't applying because they perceive academia as a career they don't want. And yet the departments don't necessarily have the money to do what needs to be done. The dean, the provost, and even the president have to be willing to redirect money because gender parity has now become a priority. We all know how fast universities implement the missing human studies requirements once the government yanks biomedical funding: one week later they are up to code. So it does get the attention at the right level, and then they have to be able to put pressure all the way down. I am surprised the provost didn't start yanking some of the dean's slush fund. That's what I would have done.

Robert L. Lichter, Dreyfus Foundation: I wanted to make a couple of comments that don't at all take away from the premise of the discussion but may help put some things into historical perspective. First, I would argue that the origin of the problem is not the grants themselves or the concept of grants but the structure of universities, because that structure long preceded the post-World War II grants-based mechanism for research support. For example, the argument doesn't apply in disciplines and departments in which grants play a much less dominant role, such as in humanities departments. I was a vice provost for research and graduate studies at a doctoral institution where you could find the same kinds of circumstances for women but where grants weren't the source. However, the grant-based mechanism can indeed exacerbate the barriers to the advancement of women.

My second point is really a side issue. A recent article in *Science* about Italian universities (including Bologna) pointed out that in the 19th century they were actually very welcoming toward women.

Debra Rolison: They were. Germany and other European countries are discussed in Noble's book.[43]

Robert Lichter: Portugal actually was hostile.

Debra Rolison: I think Portugal still is. I know a number of women scientists from Portugal.

Catherine Woytowicz, American Chemical Society: For the few of you who stopped me in the hallway and thought that I was complaining about the ACS, I want to say that that is absolutely, 100 percent, totally wrong. The ACS is the best place I have ever worked; I think it is a fantastic environment, a fantastic resource. I cannot say enough about it. If you are not involved, you should be. Really, I cannot say enough about it. It is what brought me here to Washington.

A second point is that while you made me feel really bad that I am an academic dropout—I quit academia to take this job—I guess I may now feel browbeaten into going back.

The third thing that I want to say is about women who are trying to make a case for their own advancement. Men are traditionally far better than women at self-promotion, and we haven't been doing a lot of that. So when you are doing your networking with those five pink pages, be sure to speak up when you have something worth crowing about. And if no one else says that you have done a good job, be sure that *you* say you've done a good job. Because it takes a few people saying, "Oh, yes, she did a good job," and if they hear it enough they will start saying it, too. That is one way that a little bit of promotion goes quite a long way.

One last thing, kind of a horror story. One of the reasons I nearly didn't go into academia was a professor—you probably know who she is, and if I said the school you would know exactly who she is—who was between departments. She had a joint appointment, which meant she was neither a bird nor a bee. She got pregnant, and with the stress of trying to make tenure in both of the departments and trying to be all these different things, she suffered two miscarriages in 2 years, while her students were working 70-hour weeks. She was put on bed rest. The whole world was going kaplooey and all of us came to the same conclusion: Why do we want to do this?

There is a lot to be said for an environment that is good for women and much to be discussed about an environment that is bad for women. If you don't have the right kinds of policies in place, go back home and see what you can do about them.

Debra Rolison: I very much agree about the self-promotion. As women we are socialized not to brag, to be modest. You don't necessarily have to do it the way the men do, but you can at least tell your colleagues what your latest neat result was. Just focus on the science. I recommend Liz Sobritsky's article in *Analytical Chemistry* on 1 April.[44] She interviewed 28 women analytical chemists, and there were a lot of these very same points brought up. The point is that you really can tell people about yourself and just focus on the science, because your colleagues aren't necessarily paying attention to what you are doing. They are busy, so you have to tell them what you are doing, and it doesn't have to look like hype or bragging.

[43]David F. Noble, *A World Without Women—The Christian Clerical Culture of Western Science* (New York: Knopf, 1992).
[44]Liz Sobritsky, *Analytical Chemistry,* 1 April 2000.

Marion C. Thurnauer, Argonne National Laboratory: Your comments reminded me of two studies carried out in the mid-1980s, one at MIT[45] and the other at Stanford.[46] These studies examined graduate student life and surveyed both male and female graduate students, asking questions about their situations as graduate students. I believe the conclusions were that male and female students had similar (rather negative) assessments of the situation. Women, however, were willing to discuss the situation and act on it—by dropping out. It was disheartening to see similar articles a year or two ago, because they appeared after a student had committed suicide. I think you have offered some solutions to this somewhat negative situation. For me, this is the difference between now and 15 years ago.

Debra Rolison: I like to be provocative.

[45]M.S. Dresselhaus, *IEEE Trans.* Educ. E-28, 196 (1985).

[46]L.T. Zappert and K. Stansbury. "In the pipeline: A comparative analysis of men and women in graduate programs in science, engineering, and medicine at Stanford University." Tech. Rep. Working Paper 20, Institute for Research on Women and Gender, Stanford University, Stanford, CA, 14 November 1984, p. 11.

7

Remarks for the Chemical Sciences Roundtable Dinner

The Honorable Eddie Bernice Johnson
U.S. House of Representatives

I am excited to be here to discuss science and technology with you. Science and technology have transformed the face of everyday life in the United States. As important as these fields are to the world today, their value will only increase in the new millennium. That is why the work of the Chemical Sciences Roundtable and the National Research Council is so important and why I am happy to join you today.

I am really very stimulated by science and technology and all that scientists have brought us. I stand here as a result of scientific research because I am a two-time cancer survivor. Without scientific knowledge, early intervention probably would not have taken place. I know that the knowledge had to come from somewhere, and it came out of people like you in this room.

We have progressed rapidly. When I came to Washington almost 8 years ago, I was told that we were in an age of technology. Then we were in the information age. Now it is the digital age. In less than a decade we have changed our titles because we have been so influenced by the rapidity with which we have had to change with the technology, and the stock market reflects that. The state of the art changes almost daily, and if you don't believe that, invest in technologies and you will find out.

What you and I have to do is influence the Congress to make investments in the future of science. Our most research-intensive industries have been growing at about twice the rate of the economy as a whole for the last two decades. I don't know where the economy would be if it were not for all of this innovation and the coming together of minds that take that research and make it into a reality that is marketable. As a recent report from the White House Office of Science and Technology Policy observes, "Technology is transforming the very basis of competition—enabling small businesses to perform high-quality design and manufacturing work that previously required the resources of big business, while allowing big businesses to achieve the speed, flexibility, and proximity to customers that were once the sole domain of smaller firms."[1] I am from an area that is in that environment, and it is so fascinating that the new CEOs don't wear these dark suits and white shirts and ties. They are in sneakers,

[1]Office of Science and Technology. 1997. *Science and Technology: Shaping the Twenty-First Century.* A report to Congress.

khakis, and T-shirts. They have no gray hair. It is absolutely amazing to see all these very rich young people who have thought of something that the rest of us are just catching on to.

Although the rapid pace of technological change can be unsettling, most people in the United States seem to appreciate the benefits of technology and have an insatiable appetite for more. Perhaps this is due in some part to the growing public belief that much of our current economic boom is being fueled by technology.

That belief is well founded. Studies indicate that as much as 50 percent of the economic growth of the United States over the past 50 years is due to technological innovations spurred by investments in R&D. Our most research-intensive industries—aerospace, chemicals, communications equipment, computers and office equipment, pharmaceuticals, scientific instruments, semiconductors, and software—have been growing at about twice the rate of the economy as a whole over the past two decades.

Even such a cautious observer as Alan Greenspan has acknowledged that investments in information technology have made today's low-inflation, high-employment economic boom possible. Only through the higher productivity generated by information technology can such growth be sustained.

On the surface, then, all would appear to be well for the future. Public support for science and technology appears to be strong, policy makers are crediting our investments in science and technology for the current economic prosperity, and even the Congress has put aside its usual partisan squabbling and agreed on the importance of science investments. So, should we have any concerns about the nation's future prosperity?

Last fall, the House Science Committee, on which I serve, heard testimony from Professor Scott Stern of MIT on a disturbing new study. Stern found, as have others before him, that our current economic prosperity is due in part to technological innovations spurred by past investments in science and technology. But looking at recent trends, Stern concluded that U.S. leadership in the future was by no means assured. As one sign of this, he pointed to the decline in the national talent pool: since the late 1980s, the nation's scientific and technical workforce has been declining as a share of the total workforce, and graduate school populations are flat or declining.

Moreover, a recent survey supported by the Sloan Foundation reports that there are now trends showing that the best and the brightest students are avoiding graduate science and engineering degree programs. Over the past 10 years, those taking the Graduate Record Exam intending to pursue science and engineering have declined by 16 percent. Among students scoring near the top—that is, over 700—the decline is even greater. The only bright spot is that high-scoring minority women are entering science and engineering in greater numbers.

A skilled workforce is the essential fuel to propel the economy and ensure a high quality of life. Many types of science and engineering jobs are among the fastest growing in the U.S. workforce. Now, however, we hear about shortages of highly skilled workers in some fields. Political pressure continues to grow for increases in visa quotas to allow more technically skilled foreign workers into the country.

The basic question is, Why are sufficient numbers of U.S. students not attracted to careers in science and engineering, particularly since the opportunities seem to be so great? I believe this is largely due to demographic trends and to the state of K-12 science education.

Historically, non-Hispanic white males have made up the predominant population group supplying U.S. scientists and engineers. According to Census Bureau projections, this segment of the workforce population will decline, from 37 percent in 1995 to 26 percent by 2050. These projections imply that this group will not provide the needed scientists and engineers, particularly since participation rates in these fields are also level or declining.

Clearly, it will be necessary to attract greater numbers of women and minorities to careers in science and engineering in order to avoid devastating consequences for the future. Some progress has been made over the past two decades, but much remains to be accomplished. On the basis of the 1980 and 1990 census figures, the percentages of women in the scientific and engineering labor force have increased in nearly all fields, although often from a low base level. Similarly, the proportion of female Ph.D.s in science and engineering has been growing. For example, before 1973 only 0.2 percent of Ph.D.s in the engineering labor force were women, but of the workforce graduating between 1983 and 1992, 9 percent were female. This is still a small number, but a big increase.

In 1995, women constituted 50 percent of the U.S. population between 18 and 30 years old; blacks accounted for 14 percent and Hispanics for 13 percent. In that same year, 46 percent of science and engineering bachelor's degrees were earned by women; 7 percent, by blacks; and 6 percent, by Hispanics. The proportions of doctorates were even smaller: 36 percent, by women; 3 percent, by blacks; and 3 percent, by Hispanics.

Further progress in increasing participation by women and underrepresented groups in science and engineering will require strengthening K-12 science education for all students, with the emphasis on *all*. Clearly, all students need a basic grounding in science and math to function in an increasingly complex world and to lead fulfilling lives. To meet the goal of ensuring a full pipeline of students moving toward careers in science and technology, it will be necessary to stimulate interest among demographic groups now underrepresented in these fields.

Most workplaces are becoming increasingly technological, while our society is becoming increasingly diverse. We run the risk of a widening gulf between those with the training to thrive in this new work environment and those lacking the basic skills to qualify for the high-tech workplace. To help address this problem, I introduced the Mathematics and Science Proficiency Partnership Act last year. I want to thank the American Chemical Society for its endorsement of the bill. It has now been incorporated into a broader science education bill, the National Science Education Act, recently introduced by Congressman Vern Ehlers.

My bill is a targeted measure. It seeks to bring schools with large populations of economically disadvantaged students and businesses together to improve science and math education and to recruit and support students in undergraduate education in science and technology fields. The components of the partnerships include support to the schools for teacher training, educational materials, and equipment. It also supports the establishment of college scholarships for promising students and on-the-job internships with industry.

The National Science Foundation will be authorized to award partnership grants. The awards will be based on how effectively the schools and businesses have forged their alliances and on the level of resources the private sector partners will provide.

Businesses participate by setting up college scholarships for promising math and science students, establishing job-site mentoring and internship programs, and donating computer software and hardware to participating schools. Schools participate by providing innovative training for their math and science teachers and informing their students of career opportunities in science and technology fields. Reform of K-12 science education is an enormous undertaking, and I recognize that my bill would make only a small contribution. I would like to conclude my remarks by commenting on what I see as a necessary component to achieve the overall goal.

In 1995, the President's Committee of Advisors on Science and Technology formed a panel to study applications of technology in K-12 education. Two years ago, the House Science Committee heard testimony on the findings of the study from David Shaw, who chaired the panel.

Two major themes that emerged from the Shaw panel's report were the inadequacy of funding for education research and the absence of activities connecting research to effective learning strategies in the classroom. The Shaw panel decried the anemic funding for education research, pointing out that less than 0.1 percent of the national K-12 education budget is allocated for research. It recommended ramping up the national investment in education research to $1.5 billion per year.

Reasonable people may disagree on the appropriate level for funding education research. But the current level of education research is grossly inadequate. We cannot hope to educate tomorrow's employees and managers without more basic research into what works in the classroom. Our technology-based society is undergoing constant change, and additional research will help us master these changes.

Another prominent recommendation of the Shaw panel was for a major federal investment in large-scale, rigorous, well-controlled empirical research aimed at determining which educational approaches are most effective in practice.

The Shaw report was a major impetus for the Administration's Interagency Education Research Initiative started last year, involving NSF, the Department of Education, and NIH. I believe this is a valuable research undertaking, but I cannot help observing that the first year's funding totaled only $28 million. This is a very small first step toward the Shaw report's call for a major research initiative and for a greatly expanded research investment.

Of course, funding is only one part of the equation for instituting an effective program of education research. And by effective, I mean research that will actually result in improved learning in classrooms across the nation. Current education research activities appear to be uncoordinated, and practitioners largely ignore the research findings. Adding money to do more of the same does not appear to be a fruitful approach.

Consequently, I am very interested in the proposal from the National Research Council for a strategic education research program. The scale and focus of the proposed program seem to be consistent with the goal of identifying the policies and practices that will lead to improved student learning in all schools. The question is whether the framework of the proposed research program is feasible and whether it will lead to the kinds of collaborations required among researchers, practitioners, funding agents, and policy makers.

The importance of K-12 education to the nation's future is without question, and the national investment of well over $300 billion per year is consistent with this importance. To derive the maximum benefit from this substantial investment, it is past time to end the apparent disconnect between educational practice and basic research on human development and learning. Educational reform will have a much greater chance to succeed if informed by quantified knowledge of what works.

The nation must take advantage of the human resource potential of all our people if we are to succeed in the international economic competition of the 21st century. This will require that reform efforts in science and math education be founded on educational materials and practices that are derived from rigorous research. We must engage and cultivate the interest of all children. Thank you very much for your attention.

DISCUSSION

Participant: I know you are doing something about education. Could you tell us a little about it?

Congresswoman Johnson: Actually the effort is about making sure that there is opportunity for more enhancement of teachers' skills and, also, trying to bridge education and industry, setting up partnerships and internships for teachers and students. We are doing a lot of this in Dallas, and that is where I got the idea, but I think it works. We have to remove the mystery surrounding industry for educators, letting them see the environment and understand it so they will know what they are preparing students for. That is one component.

The other is simply to strengthen the curriculum and make it sounder and to strengthen the skills of our teachers, building partnerships, opening dialogues, integrating these various skills so that our young people will know what it is like to be a scientist in a lab or a scientist in industry and know what that environment is. The important thing is to be able to read about, think about, and understand intricacies that we never thought about in the past. Think about the better investment of time. I remember when I was growing up—and this will tell my age—we talked about the long hand and the short hand to tell time. It has been many years since we talked about that because we don't have to do that anymore. You just look at your watch, and it tells you what time it is, but that is where we have to leap. We have to go with our young people so that we can close that gap and use that valuable time for something more updated.

Participant: With regard to young scientists, what would you do?

Congresswoman Johnson: That is kind of unfair in a sense since I am on this side, but if I were you, I would write letters. I would visit my legislative representative periodically. I would also include students and bring students with me to further that dialogue. Even when the legislator cannot understand all of the science, he or shc can understand enough to know that things really are changing and that we have got to let that change happen, and it has got to happen with our support.

I remember I put a statement similar to this on several pieces of legislation, but one of the ones that I remember so clearly was telecommunications reform. It seemed so simple at the time, and I said that I understood it enough to know that I shouldn't be against it. I couldn't explain why I was for it, but I knew enough to know that I couldn't stand in the way, and that is really about where we are in terms of moving forward. We don't know what else we can learn from the universe, but we know that we must not stand in the way of exploring that universe.

Participant: There was a resolution passed last fall specifically to address the issue of women and minorities in science. Can you tell us a little bit about where that is and what progress has been made?

Congresswoman Johnson: You know we have always had committees like this, but we have never had to bring results and recommendations, and that is the one thing we added this time: to bring not just a report but also recommendations, so that we could try to address them. We are getting that report, I think, in about 2 weeks. I hope some of you have been involved in some of that.

Richard C. Alkire, University of Illinois at Urbana-Champaign: I know the future is always hard to predict, but we have before us the President's budget, which was one of the most remarkable budgets for science in decades, and I wonder if you could comment on the sense of what might occur?

Congresswoman Johnson: In terms of the authorizing committee you are in good shape. Now, in terms of the appropriators, that is the next story. It is somewhat hard to predict, although I think each of us on the committee has done our own private lobbying to try to convey how important it is, and that we realize how difficult it is to address.

When you are on that committee you hear so many stories and every issue becomes important. You just never have enough money for all of it, and that is where the bottom line will be. I think that you can help with that. The appropriators are highly responsible people, but—as you know—they are subject to influence like anyone else. I hope all of you have written about the importance of scientific research, or called someone, or e-mailed somebody, but if you haven't done so, get the list of who is on those committees and start to influence them. It will make a difference. They need to understand.

It is so interesting: probably the strongest lobby in Washington is AARP. They don't have a PAC, but they have all the time in the world.

Keep this in mind. Try to do what AARP does. Keep the issues on the minds of those appropriators, because they have the final say-so. The President has really taken the lead in emphasizing the need for more scientific research and for more dollars going in that direction. So, he has taken the lead. Many of us help to spread the word; the push in that direction was a bipartisan effort. The Science Committee has supported that, but we only can authorize. We don't appropriate. So, it is the appropriations process that we must complete before we go home. That is the only really essential thing we have to do before we leave, and that is where you can help them understand the importance of keeping our country on course and keeping this prosperity going. The only way we can do that is with the help of people like you, and we cannot do it standing still.

8

Reports from the Breakout Sessions

Following the presentations described in Chapters 5 to 7, breakout sessions were organized to enable more extensive discussions among the workshop participants. The following questions and statements were suggested to the breakout groups as possible topics for discussion:

- What are the negatives and positives of formal diversity programs?
- Compare employment practices in industry and academia.
- Are the best practices of industry transferable to academia?

Discussion leaders from the breakout sessions then reported in plenary session what they believed to be important ideas and topics that had emerged in the discussions.

Frankie K. Wood-Black, Phillips Petroleum: Our group began by talking about formal and informal diversity programs, and we discussed problem solving for some recruiting issues that had been observed by the participants.

We started with mentoring programs, both formal and informal, trying to identify some of the pros and cons—what worked and what didn't work. One of the characteristics of the programs that didn't seem to work was a high degree of structuring, with a sense that the mentor-protégé relationship was assigned rather than chosen. There was no flexibility for participants, no room for volunteering; it was, "You will be so-and-so's mentor," and that approach didn't seem to work very well.

An example of a program that seemed to work very, very well was GE's mentoring program, in which the company had 600 executives who were told that they each needed to find a younger mentor to teach them about the Internet. So it was highly focused from the standpoint of being an assignment, but there was flexibility in how the relationship was developed. It actually worked quite well.

We have a program at Phillips, for folks who come into the plant, yard men included—folks who dig the holes and turn the wrenches. It has a focus on safety, and it is a peer mentoring program, but it is voluntary and seems to work fairly well. The structured formal programs with mentor assignments were not so successful; some programs worked and some didn't.

We talked about recruiting these employees from the perspectives of a selection committee and an industry recruiting committee. One of the common patterns that emerged is that the successful recruiting committees were proactive, they weren't reactive. That proactiveness seemed to help make those recruiting committees work.

We also noticed that there were champions: successful candidates who came out of the recruiting committee process had a champion who said, "Did you see this one?" They had somebody to speak up for them. So champions were very important.

Finally, we talked about some of the diversity programs—seminars, talks, and the like. One of the programs that was mentioned was Pat Heim's "She Said/He Said" program,[1] which talked about communication styles.

This is where we started bringing into our discussions the question of problem solving for employment procedures. We talked about hiring strategies and, particularly, the differences between academia and industry. Industry's hiring practices are more formalized—there are procedures and people know how they work. There are some trends that ebb and flow—they may be cyclic—such as going from a central corporate environment to a business-unit-type environment.

What is happening in hiring practices is very similar. Some industries are focusing on recruiting from a core group of universities, while others are branching out once again to get more diversity in their hiring pools. One thing that did come out for the academic institutions was the existence of human resources (HR) departments. The fact that corporations used HR department structures did seem to help the recruiting process, whereas academic institutions frequently must develop these structures every time they hire a professor. While HR departments aren't the solution, they provide a structure. They are going out and doing the same things every time, so search committees do not have to recreate procedures every time. Academic institutions are at a disadvantage there.

What are the problems in recruiting, and what are the effects of locations—where the candidate visits during recruiting—and culture perception? Once a candidate submits a job application, what happens next? "First perceptions are killers," was a recurring theme. There is something about the first perception, when a candidate is out in the organization or in the department for the first time, that can be a killer. Even though everybody is trying to be proactive, something frequently happens to cause a negative perception during these visits. The question came up, Is it a hostile environment? How do you define a hostile environment, and how do you address a hostile environment? These are significant questions.

We also talked about flexibility in hiring. If you go looking for the person you think you need, sometimes you may be defining that person too narrowly. You may be saying, "I need a biochemist in XYZ area, and the candidate needs to be a female, needs to be a minority." You have actually defined that person so narrowly that you may not be able to find that person.

Finally, one difference between industry and academia emerged—and this shows why human relations departments may be good things. A lot of industry hiring practices are structured in response to federal grants and take into account that they will be audited against a particular regulation. For example, you may generate a candidate pool, but the auditor comes in and asks what you specifically did to ensure that a woman, a minority, a Vietnam veteran, or any one of a whole host of other candidates was listed. What did you do that would let you say you went that extra step in recruiting for the position?

We talked about different approaches to the equal opportunity philosophy. In industry, a lot of these approaches have been driven by knowledge of what the auditor is going to look for. When you are in an academic environment, you don't necessarily have that structure.

---

[1]Pat Heim, "She Said/He Said: Gender Differences in the Work Setting," audiocassette, The Heim Group, 1995.

Industry uses benchmarking for everything: costs, supply chain, all those kinds of things. Maybe that is where people need to go in hiring: to benchmark departments in terms of what is or is not a good environment and to look for the blind spots. We heard from folks who have a proactive department; they have a critical mass of women, yet it is still perceived to be a hostile environment. Where are the blind spots? Benchmarking should be able to help.

We talked a lot about goals. What is realistic? What is a stretch? Are goals good? Markets measure all those kinds of things. A running theme was, You've got to keep pushing, you can't just stay unchanged.

Lou Ann Heimbrook, Lucent Technologies, Bell Labs: Let me share with you a little bit of what came up in our group discussion of formal diversity programs. We discussed the positives and negatives for formal diversity programs, whether those programs described were in industry or in academics.

The first positive is financial incentives. These may be financial incentives for the diverse candidate or for the person bringing in the candidate. Often it is the formal programs that force that. You have to watch out for the flip (or negative) side of that as well.

Formal diversity programs can broaden the notion of what the "best" is. What we are trying to do is entertain the thought that the best and how we define it can change and become much broader. Diversity programs bring that to the forefront for individuals.

Diversity programs also can provide formal training. Where formal diversity programs are in place, all aspects of managing diversity typically can be handled. We discussed some courses that have been organized, such as "Men and Women in the Work Environment." Programs at the graduate student level appear to be working well in the academic arena. That seems to be quite positive.

Formal diversity programs also can expand comfort zones, to make the work environment more sensitive to cultural differences. At the very least, they can raise the consciousness level. These are some of the positive aspects that we thought the programs would bring.

What about the negatives? As you can imagine, if you are using financial incentives to change the workforce, there is already a workforce in place. The incentives can aggravate the financial differentials between the haves and have-nots. So you need to be quite sensitive, because diversity programs can drive differentials.

In addition, diversity programs may cause major institutional change or disruption. Formal diversity programs often drive change, whether it is in a university or in an industry. That might be viewed as either a positive or a negative, but remember, any workforce will have both components.

One thing that I want to emphasize, which was mentioned quite often, is the feeling that there is little leverage to require formal training for faculty. One perception was that this might be seen as infringing on the academic freedom of faculty. However, this clearly is not the case in industry, where formal training already exists. So this is an issue that academics would have to address in their formal programs.

Now, how would that kind of training influence employment practices? Let's start by taking a look at employment practices from an academic perspective, and then we'll look at industry and other sectors. One of the employment practices is the old-boy network. What happens typically in the hiring practice is, "We have a chemistry position, let me call so-and-so, so-and-so, and so-and-so." This is an employment practice now in use. In this audience, I am now looking at what I hope will become the old- or young-girl network or the women's network. The other employment practice we discussed is the manila envelope, as often used by formal search committees in academia. They get the envelopes in and then they sort them into "desirable" and "undesirable."

Advertising job openings is a common requirement. There is variation in the length of time the opening must be advertised, but as long as you advertise, you can say you have followed the employment practice. Employment practices may vary by institution, just as they would in industry. Advertised positions require letters of recommendation or information on the candidate, so a portfolio can be made for the candidate; this provides the maximum amount of information.

Employment practices may include some discussion of employee needs, but typically there is minimal consideration of spousal accommodation. We have alluded to the question of dual careers, and this is often discussed. There can be a lot of spousal professional support. Typically in industry, when someone is being interviewed, everything possible will be done to make sure the portfolio includes the spouse by, for example, helping them find employment.

But practices sometimes lead to legal violations in interviews. Some of the questions asked during the interview process are illegal, but this does occur. Those involved in the process need to recognize the importance of absolute adherence to the letter of the law—knowing that you can be sued if you fail in this. Discussions need to be limited to what is relevant to the job. They should not extend to anything outside the job, the job description, and the skill set necessary for that job.

The positive practices being developed in industry include early recruiting or outreach to preprofessionals. The goal is to contact potential workers before they are ready to enter into your industry. There is active attention to diversity; it makes economic sense, so there is active attention to it in most cases. We cannot assume that the candidate wants to join the company, so we must sell the company. One of the things that we are looking at is an in-your-face type of selling, a selling attitude toward employment. But I like this quote from one of the individuals in our group: "I just want to make sure they are not too close to your face."

In other institutions such as the national labs, our group noted the excellent employment practice of screening postdocs for their suitability as future staff. I know this is difficult in academia, because you typically have one slot, and you may not want to hire someone from an academic postdoctoral position to do the same organic chemistry as the mentor. But screening is something that might be useful.

What aspects of best practices can be moved from industry to academics? One of the overarching things is positive reinforcement for enhancing diversity. However you choose to measure, or whatever your metric is—whether it is monetary or perhaps lab space—positive reinforcement for that tends to be something industry does well. Finding mechanisms to transfer this to academics would be useful.

There is also the redirection of emphasis from individual achievements to team-oriented or departmental accomplishments. With that redirection, you automatically increase the likelihood of diversity because your team has just become larger. A move from individual to team is one way to do this. I also believe the idea of selling can be transferred from industry. Sell your institution to the candidate. If you want to bring a top candidate into your institution, make sure you sell it. Don't just assume the candidate will want to come. And finally, there is the need for more explicit mentoring and guidance. This exists as both a defined and an undefined process in industry.

I would like to leave you with a couple of key points that I took from our discussions. The first one applies to academics: you seem to get it for students, but not for faculty. What we consistently heard is that the approach and principles used for graduate students for both recruiting and diversity—seem to be excellent across the board, if you look at the graduate student profiles. How do we apply the same approach and principles to recruiting faculty? When you take a look at that, it becomes apparent that diversity isn't a given. In both industry and academia it is used when it makes good business sense. The reason that we achieve such diversity on the graduate student level is that we have cheap labor. That is why you want them in there to carry out whatever programs you have.

On the business side, we have quite a diverse set of customer portfolios, so it tends to be a money-driven activity. How do we address that? Overall, we need to be quite proactive in addressing any sort of workforce issue, because that lets us get a jump on the competition, whether it is another institution or an industry.

Geraldine L. Richmond, University of Oregon: Ours is the soul-searching group. After Debra Rolison's talk, we spent a lot of time discussing academic institutions as a hostile environment.

We talked about differences between industry and academia, particularly the value systems. In industry, there are clearly metrics for success and a better idea of what success means. The success of employees is more closely tied to the success of the company. In academia, the metrics for success are not as clear. For example, an academic institution may be considered successful if it keeps the student body numbers up. This provides the right number of undergraduates, who pay tuition to keep the institution going. But faculty salary increases and faculty promotions are not tied to that, except for institutions where the faculty spend most of their time doing undergraduate recruiting.

The metrics for success are not as clear in academia, and the reward structure for faculty members varies from one institution to another. The reward system is changing all the time. Academia could learn from industry on a number of issues. It is not necessarily true that all industry is more progressive in the area of child care, but there are good examples of companies that have progressive policies toward child care and family—more so than many academic institutions. There are also in industry examples of healthier workplace policies, particularly on the issue of pregnancy in the workplace. We have graduate students, undergraduates, and faculty in academic institutions who are in their childbearing years and have children. Yet the issue in most departments is largely ignored, and little effort is expended on setting policy for pregnancy in the workplace. Academic institutions could learn a lot from industry on this issue. Management training is also more prevalent in industry—in fact, it is almost nonexistent in academia. I said almost, because management training in academia usually consists of watching somebody else.

Industry appears to be more family friendly, so the question became, Why? One issue that was raised had to do with retention. Retention is very important in industry. In academia, if a faculty member leaves, you soon have a pile of 200 applications sitting next to you to fill that one job slot. So retention takes on a different character in academia. You want to retain someone if they bring in a lot of money and have national recognition or if they bring something important to the institution—but that may not come until later in their career. In industry, there is a greater push to retain people once they walk in the door. This leads to policies that are more supportive of the community on issues—child care is one—that have been on the back burner for many academic institutions.

Let me now go to what we spent the majority of our time discussing: the hostile environment in academia and other issues. There are mixed feelings about how hostile the environment is in academia. There are some very positive situations in academia, but many others are negative.

One situation in academic institutions that is perceived to be particularly negative is that of the very large research groups that depersonalize a graduate student's experience. These are research groups where students just turn the crank and spew out results but have very little contact with the other students in the group. That can be a negative experience, particularly for women. We see funding being directed more and more toward large group activities, which causes us much concern, not just because this means that one PI must support a huge group, but also because in mega groups, the student becomes a cog in a big machine.

I think it is important to point out my own perspective on NSF's tendency to treat training and the training experience as a valued commodity in the review process for grants. I believe this will actually help to maintain the strong incentive to have faculty members stay tied to their students, and that is a valuable result.

There was a discussion on small colleges versus universities: Which is better, which is worse? Which is more family friendly or unfriendly? The perception is that small college environments offer a more flexible and supportive environment, but that is not always true. There is insufficient time to discuss this in detail, but students need to get a realistic picture of what life is like at a small college compared with a larger institution. There are tremendous demands on women at small colleges, demands that often are much more time-consuming than at a larger university.

Finally, the expectations for academic faculty are rising everywhere, contributing to what might be perceived to be a hostile environment. Technology, travel, and information overload are all contributors. That is not just in academia, but also in industry.

At small colleges, we see increased pressure on faculty to have a research program, to publish, and to get research money. The metrics are changing all the time, and that just places larger demands on faculty at small colleges. At larger institutions, faculty are required to obtain multiple grants, whereas their predecessor or their mentor might have been able to survive with just one. Collaborations are now called for in addition to individual programs, and many faculty are asked to be involved in entrepreneurial activities and start their own small businesses. All of this means that there is a tremendous amount of pressure in the academic community, contributing to a hostile environment.

So what is my overall message from this discussion? There is clearly a lot of hand-wringing and heart-wrenching and soul-searching going on about academia and the hostile environment, at least the perceived hostile environment in academia. I think the message that needs to be given to women who are going into academic positions, and frequently industrial positions as well, is the need to set priorities—the need to set one's own personal priorities.

We are hearing that women want everything and want it all and want it now, and we are hearing that from the younger women coming up the career ladder. Unless women learn how to set priorities and reevaluate these priorities as they go through their careers, I'm afraid that we're asking them to follow everybody else's priorities, which may be overwhelming. The demands placed on us in the workplace are large, and they are growing. These demands encompass family issues, teaching, and research, as well as entrepreneurial pressures. If we don't set priorities, we'll just wind up feeling that the environment is even more hostile than it actually is.

W. Sue Shafer, University of California, San Francisco: We began our discussion with some legalistic questions about what formal diversity programs are. Obviously, they take a couple of different forms, including those that are designed to improve the diversity of the workforce and those that are designed to improve the sensitivity of others in the organization to diversity issues.

Someone in our group asked if you can you have quotas these days. The circumstances in which it is legal to have quotas are those in which you can document that you are addressing previous discrimination. Often quotas are unspoken but visible. Then we turned to small companies and discussed what size businesses must be concerned with diversity programs. Some of the new small start-up companies may be below the threshold size for which the programs are required. But we can all think of other reasons that may drive people to have such programs, even if they are not legally required to have them.

Our group included people from academia, government—both administrative and laboratory environments—and industry, so we were trying to synthesize across all of these groups. One of the people from one of the government labs reported that they were doing quite a bit of diversity training. Some people in these programs are worried that if they don't accept everything taught in the diversity programs, it may somehow come back to haunt them, and they will be fired. I don't know that we have had any cases of that happening, but at least the concern was expressed. We talked about the consequences of some of the diversity training that goes on. In one industry setting, the goal was to have zero so-called "personnel incidents." Now that is obviously not attainable, but it is certainly an interesting goal.

We discussed the fact that one young woman decided to hold some diversity training for the technicians and graduate students in her group, and some of her faculty colleagues agreed to participate as well. That was both positive and negative. One of the topics was sexual harassment, and she ran into a couple of faculty members repeating one or another version of the tired old banality, "Well, if they would come to me I could tell them how to harass somebody sexually." The training had missed its mark for those faculty.

We had a long discussion on how to get faculty to take mandated training. If this is to be successful, the faculty has to impose that on themselves. Even then, you probably will miss the 5 percent who don't want to deal with things like that. We then went on to talk about what kind of teeth can be employed to reinforce diversity training. One individual reported that somebody brought some pornographic tapes or CDs to work and left them on their desk. The employee was told by a supervisor to put them out of sightæthey were inappropriate in the workplace. That didn't happen, and the employee was fired on the spot. That was in an industrial setting. I think many of us worry that one morning we could be fired, so a clearly stated policy and consequences might be helpful in conjunction with both the training and actions. But most people feel that some behaviors—like carrying firearms into the workplace—merit being dismissed on the spot no matter what (unless you happen to work for a police department).

We talked about the difference it makes when managers absolutely believe that the diversity training they are giving should have real consequences for their organization. If top management is convinced that this is in the best interests of the organization, you will have a much better effect than if management is only going through the motions, or is doing it for some legalistic reason.

NIH has been a positive influence in stimulating training in the responsible conduct of science. A question was raised whether NIH has any requirements for training in diversity, and it was reported that it does not have such training requirements. However, NIH requires institutions to have programs to ensure equal opportunity and to have sexual harassment policies, and so on. Unfortunately, we do not have mandates for how to do it; it just has to be effective.

We talked a little bit about programs to provide incentives to hire minorities. If you recommend a minority candidate for an opening in that institution, you might get a reward if that person is hired. But there are undesirable consequences if you perceive that you were hired only because you are a woman or only because you are a minority. Finally, being the first woman or minority candidate to join a group might have significant consequences for retention. A person working in an already diverse workforce is probably better off than someone who is the first addition to a previously monolithic workforce.

We did talk about transfer of good practices from one sector to another, and we talked about the incentive programs for hiring minorities. I believe that the federal Department of Energy laboratories have used that pretty successfully, but I don't know that academia has used that at the professorial level. In our group, we didn't hear of any reports that this had been done in the industries they represented. We talked about giving departments an additional faculty position if they hired a minority candidate, and the pros and cons of that approach. Certainly NIH has been known to reward institutions that do well in recruiting underrepresented minorities. In some technical fields other than chemistry—for example, biochemisty—this has been less of a problem for women. There are examples of departments that have been rewarded with an extra position if they have a good prior record of minority recruitment.

We also discussed the dual career problem and what companies do about it. In some reports from the academic sector, the whole family was recruited to make sure that the spouse—as well as the candidate—was actually on board. You are not going to convince the candidate you want unless you convince the spouse. Industry seems to have done a 180-degree turn. They used to think that employing two people in the same company was a problem. I think they now see it in many cases as a plus. And we heard some reports of academic experiments in which positions that were initially shared became two full-time, tenured slots. That could be a way of retaining individuals in parts of the country where it may be hard to keep faculty.

I'm going to return to a theme that I have spoken about before in this meeting. I think that any institution that wants to keep and attract good faculty needs to view faculty or scientists as an investment and to treat them as a resource whose success is going to make the company or the institution better. To help that happen, I think we need the kind of diversity training that allows us to "walk in the other guy's shoes," and we can get professional help in doing that. To the degree that we can give power and control to our employees who are actually doing the work—but still be a supervisor, a consultant, and a sister—I think we are going to ensure better investments in the people that we hire.

Maria K. Burka, National Science Foundation: Our group is the one that was primarily a chemical engineering group, although there were some chemists in the group. In our discussion of formal diversity programs, we looked at some of the pros and cons. Some of the pros that came up include the observation that students flock to places that have set programs, particularly when these programs are well run. That leads into the second item, which is that you can depend on these programs to bring in a pool of good candidates. So it helps the institution as well as the students. You have students who feel confident that they are going to a good place, and it helps the institution because you are going to attract some good students. In addition, it is the right thing to do. While that may sound hokey, it is still very important.

One of the negative aspects of any diversity program is that it is an entitlement program, and then people don't always perform. A more important problem is the feeling that "This is being dealt with over there," so the rest of the people in either the company or the department don't feel that it is their responsibility. If there is a formal program elsewhere in the organization, individuals may conclude, "I don't need to get involved in this; it has already been handled."

Diversity programs work only if they are properly supervised. For example, there is a formal program at Berkeley, where they actually see candidates through from the beginning to the end. This ensures the success of the people who come in under the formal programs. There is still concern over the possible attitude of others, that you have hired a person for the wrong reason—that the person is not necessarily qualified for the job. But that works both ways. Some in our group suggested that this is irrelevant. If you come in as a new hire—no matter what reason other people say you were hired for—if you perform, you really shouldn't care. That view suggests that such concern is not necessarily a negative; it is just one of those things that is often mentioned but really is not that important.

We then talked about the various practices in industry and academia. It was mentioned that big companies—those that have devoted substantial efforts to these issues—are often bastions of diversity. But smaller companies can be the exact opposite. Very often, students know that there are certain small companies that they would never want to go near, because the environment is much worse than what they have already experienced in some of these academic departments where there are problems. Other participants pointed out that while in industry the situation is good, it is still not a utopia. We still have a way to go.

We talked about some of the specific rotation programs or field programs that are very positive. These permit a person to be rotated to different parts of a company to get good training and the background needed to move up the corporate ladder. Also valuable are brown bag programs that bring together people who have similar problems and similar interests.

Then we talked about the important idea that "what gets measured gets done." Companies are very good at measuring things, so things get done. A very positive example is evaluation of the success of subordinates as a formal part of a supervisor's performance appraisal. This is very important, because it means that the supervisor will invest effort and time into developing the younger person.

We discussed the differences within companies, that is, in different parts of the company. The plant environment—which, of course, for chemical engineers is very important—is frankly a much more negative place for women than the office environment. Somehow, the plant environment, when you are dealing with operators and so on, is not as welcoming to women as, for example, the typical business environment.

What about the practices in academia? First of all, the programs are much less formal than they are in industry, but some of these things are changing. Certain departments now have what they call post-tenure review. How far will these go? Some schools are talking about doing away with tenure, but we don't really think that will happen. I think what happens is that these post-tenure reviews basically affect a person's salary increase. We also touched on NIH training grants, which really can be used as a mechanism for change, and finally we noted that some schools now have ethics courses, which sensitize people to some of these problems.

Some practices are transferable from one environment to another. For example, mentoring needs to be recognized as a sanctioning role. One of the things that was mentioned—and this is a situation in both industry and academia—is that very often women, particularly senior women, become sources of comfort—people to whom others talk about issues. This becomes very time-consuming, and it affects the ability of these women to do their job. There is a need to understand that mentoring is something that is positive but that the time consumed should be credited in that person's performance appraisal. What is very good in industry is the recognition that the company is making an investment in the person. Maybe academic departments should look into that approach rather than bringing in young faculty and letting them sink or swim. Early feedback—particularly to young faculty members—would help. Employees in industry get annual or semiannual reviews; you don't necessarily get those very in-depth reviews in academic institutions. The issue of mandatory mentoring of young employees—which was mentioned by another group—has been instituted in some places, but this very often doesn't work, because you are forcing something that is just not there.

Teaching management skills to department chairs and deans would help. Very often, the people who become department heads or deans are appointed because they are first-rate researchers. Often they don't have the people skills and they don't have the management skills. Management skills could make a big difference with some of these diversity programs. In academia, each faculty member is an individual little CEO. Each group is like a company, with the faculty member as head and their research students as workers. They have to bring in their own funds, they have to support their own group. Yet, they haven't been taught the skills to be these little individual CEOs. Maybe that is something that individual institutions could work on: How do you deal with your students? How do you help them? How do you promote them? How do you get them on their way? How do you bring in money? And so on. How do you manage the operation?

These mentoring and management issues are connected. It may not be possible to directly transfer practices from industry to academia, because these are different worlds. The reputation of a particular department—which may consist of anywhere from 10 to 40 people—is dependent on the people it brings in. A company with hundreds of thousands of personnel is operating with that same one-on-one personal situation. When you are interviewing someone for an industrial position, you are not evaluating how that person's performance will make your department's reputation. But academic departments are looking to raise their reputation.

At the same time, that person in a small group is going to live next door. You want a person with whom you are going to talk, with whom you may play basketball, or whatever. You don't necessarily have that in a company setting, because you are hiring someone who could be moved to different parts of the company. So some of these practices are not necessarily transferable. When interviewing a potential new faculty member, people often want someone who looks just like themselves, because the new person is going to live next door.

If I have one summarizing statement, it is the word accountability. I think that is the bottom line in all of this. Universities basically are not held accountable for hiring, and they are not held accountable for the atmosphere; but very often industrial organizations are. Companies talk about the atmosphere; very seldom do people talk about atmosphere in an academic environment. They talk about the department's reputation rather than the atmosphere in the department—but atmosphere should be taken more seriously.

9

Experience of Women at the Massachusetts Institute of Technology

Nancy H. Hopkins
Massachusetts Institute of Technology

It has been nearly 6 years since I became involved in an extremely interesting study at MIT that looked into the lives of the tenured women faculty in the six departments that make up our School of Science.[1]

This study was conducted by the tenured women faculty in collaboration with a few high-level administrators and with the support of the president of MIT. From this study, and from responses to it, we learned that the playing field for women in science, even for those who are extremely successful, is not a level one.

What is interesting about the MIT study is that when we began, most of us thought we were dealing with a very unusual situation. We thought we were dealing with the problems of an unusual group of women in an unusual institution, women who were striving for an unusually high level of achievement in an area that few people ever participate in. I recall one of my colleagues saying, "Nancy, you're wasting your time. No one, not even other women, will ever care about the problems of these few women scientists."

But to our surprise, last year we learned that we were in fact dealing with a very widespread, if not universal, problem for women in science, and indeed in many other workplace settings, including law firms, the military, businesses, the arts, and so on. Not only that, we were definitely not the first people to figure this out!

Today I would like to do three things. First, give you a brief personal account of how I came to be involved in the MIT study and what we learned. Then I want to tell you about what happened after our report became public, and particularly what is happening at MIT in the aftermath of this study. Then I would like to make some suggestions about how we might fix the problems that we all know contribute to driving women out of science.

[1]"A Study of the Status of Women Faculty in Science at MIT," *MIT Faculty Newsletter,* March 1999. Available online at <http://web.mit.edu/fnl/women/women.html>. This article summarizes the findings of a 150-page unpublished report on the same subject prepared in 1994.

I joined the faculty at MIT as an assistant professor. I believed that civil rights and affirmative action had solved gender discrimination and that I would not encounter it in my lifetime. I thought the only reason there were so few women on the science faculty at Harvard, where I had been a student, or MIT was that women had children and remained their primary caretakers and that men who did the type of science I was trying to do worked six or seven days a week, 10 to 12 hours a day. So it was pretty obvious why there weren't any women there; how could you do these two full-time jobs at once?

Over the next 15 years, as I served on the faculty, I found out to my surprise that gender discrimination still existed. The way I discovered this was by watching how other women were treated. Because there were so few women, it took a long time. Being a scientist, I needed convincing evidence. But finally, after 15 years of observation, I knew that women of equal scientific ability and accomplishment were not valued and respected as highly as their male colleagues.

I cannot tell you how demoralizing this was. By the time I was convinced of it, I wished that I could age faster, so that I could retire, because it was so discouraging to see these brilliant and highly successful scientists treated unfairly. What kept me going was my passion for science—and the fact that I thought I was the one exception. I had a very difficult life at MIT, but I did not see that the same thing happening to these other women was also happening to me. I explained away each unpleasant incident by its special circumstances.

Looking back on that period of my life, I have to conclude that my failure to understand what was happening was due in large part to denial. I am a big proponent of denial, by the way.

But beginning about 8 or 9 years ago, a series of events occurred that opened my eyes. I wanted to change my research direction, and I needed to get some resources from the university. These were very modest resources—a small amount of additional lab space, a modest piece of equipment—things that everybody else already had had provided to them in quantity. At first my administrator helped me, but after another one stepped in I found it was extraordinarily difficult to get the things I needed. One day, a woman who washed glassware for the labs in the building said to me, "Nancy, why do these men have so much and you have so little?" It was that obvious. I should say that this struggle took about 50 percent of my time and 90 percent of my energy. Every day I would go home and try to recover from that day in order to prepare for the next one.

Finally, after some years of this, there was an incident that proved to be the last straw. The one that led me to say, "That's it." The day it dawns on you that for all those years it is possible nobody ever saw you as an equal in the profession to which you gave so much of your life—that day is a very devastating day. I felt I had wasted 20 years of my life. For several days, I was paralyzed. But then, happily, despair turned to anger. I decided I would try to solve my problems, try to change my working environment.

At first I got absolutely nowhere. Soon I had worked my way right up the administration to the level of the president. So I sat down to give MIT its last chance. I wrote, "Dear President, there is discrimination here and you really ought to do something about it." I showed this letter to a friend of mine, and he said, "You're not planning to send that, are you?"

So I thought I would ask another woman faculty member to read the letter and delete anything that might be offensive. I picked out a woman whom I admired enormously, although I barely knew her. She was enormously successful scientifically and politically correct to boot.

Looking back, it is hard to remember how difficult it was to show this woman my letter, because we have come such a long way in these past 6 years. But then I had to steel myself emotionally. I think the reason it was so difficult is because we grow up believing that if you are really good enough, you can make it on your own. Even in the face of discrimination. It had taken me 15 years to discover that it is not that simple at all. I had seen that discrimination can be extremely costly. It can prevent people from rising to the top and it can even prevent them from being recognized when they do. But I thought I was the only person who had figured this out.

We were sitting in Rebecca's Cafe in Kendall Square, a very noisy place at lunchtime. This exceptional woman scientist was reading my letter, and I was waiting to see her reaction. Her face did not change as she read. She looked extremely serious. When she got to the bottom of the letter, she laid it down on the table and said, "I'd like to sign this letter, and I think we should go and see the president. I've believed for a long time that tenured women faculty here are not treated equally."

I was speechless for a few moments. She had figured this out too! It wasn't me after all. It was really true. It was her response that was the beginning of everything that was to happen.

It soon occurred to us that if we agreed about this, perhaps other women would too. That was when we made our second surprising discovery. We, and a third woman we soon recruited to our task, got out a catalog of the faculty in order to make a list of the tenured women faculty in the six departments in the School of Science. We wrote down the names, and when we had finished, we found there were only 15 tenured women in the six departments in the School of Science, versus 194 tenured men. I said, "No, you know that is not possible, just go back and do it again."

We found two more women who had primary appointments in the School of Engineering and joint appointments to Science. We thought we had better include them too, just to get a bigger sample!

We divided this list in half, and off we went to see the other tenured women. Once again it was extremely embarrassing, because the women we were approaching seemed from a distance to be so successful. They were forever winning this prize, that award, or being elected to this or to that. But before the second sentence was out of my mouth, the first one said, "Do you have anything I could sign?" By the end of the first day, we had 10 signatures of people who wanted to sign on to our effort. We were amazed.

In the end, 16 of the 17 women decided to join this initiative. I should say, there was not a uniform opinion. Of the 17 women, there was one woman who said that she had never seen or experienced discrimination in her scientific career, and she did not sign on. Among the other 16, I think there were about 10 who had almost identical views. From the time we started, these women could finish each other's sentences. But among the others, there was a range of views. There was one woman who said she hadn't experienced the problems, but she knew that others had, and she signed on for that reason. There was one who said that her life was very difficult, but she wasn't sure if it was for this reason. She thought it was due to the many difficult people she kept running into. But she signed on because she said she had to do something or she was going to quit.

Of all the things that happened at MIT, none has meant as much to me as the formation of this group of women faculty. It is impossible for me to say enough good things about these particular individuals, their brilliance, and their integrity. These women had devoted their careers to being successful scientists. To take this step was hard—embarrassing and awkward. The women wanted to operate as quietly as possible. We wanted to fix things and just go back to our labs and classrooms. I believe these women operated out of a sense of obligation. Students would sometimes say to them, "I don't want to be like you." Who could blame them? One woman said, "I don't want to be like me, either." I believe that many of these women took action for the next generation of women scientists more than for themselves.

In any case, I must say that what was to happen at MIT, the ability to take on a powerful institution and get its attention to look at an unpopular problem probably required this solidarity. The power was in the group. When a single woman takes this issue to a powerful administrator in her institution, often the administrator simply doesn't know what to make of her complaint. Together we had a chance to make a convincing case.

We wrote a letter to the Dean of Science. "We believe that unequal treatment of women who come to MIT makes it more difficult for them to succeed, causes them to be accorded less recognition when they do, and contributes so substantially to a poor quality of life that these women can actually become negative role models for younger women."

We asked the dean to address the problem by establishing a committee that would look into the entire matter. We wanted to document the problem for the administration, so that when a woman had a problem, instead of having to go individually and prove the case over and over, there would be a body of documentation. It would help administrators to understand the problem and help each woman to explain her own case on her own. We also wanted to look at data to see whether the perception that things were unfair was really correct. Would the data also show discrepancies in terms of things like resources and compensation? Most important, we wanted to work collaboratively with MIT to study and then try to solve a problem.

But the question was, How would the administration respond? That proved to be the second key event in the whole affair. We went to see the dean, Robert Birgeneau. It was a very tense moment. But to our relief, from the beginning he was extremely supportive of our request. It turned out, of course, that it wasn't quite that simple. There was a lot of opposition from other administrators. We learned that the dean went to the president for advice. The president told him, "Just do it!"

With that, a committee was established. It consisted of a tenured woman from every department and three men who were or had been department heads. These men turned out to be critically important to our efforts. They came from the system. They were powerful people, and when they learned the whole story, they became our advocates. They went back to the system and said, "It's true, and you need to do something about this."

So what did the committee do? What we did was interview the tenured women faculty, the untenured women faculty, and the department chairmen. We also collected data on the distribution of resources for research, on rewards and compensation, and on the inclusion of women on important committees. Then together we put the data out on the table and looked at it.

The process took 2 years. The report that came out of it was about 150 pages of single-spaced 10-point type, lots of data tables, and so on. So what did we learn from this enormous amount of work? From interviews, we found out the following: We found that young women faculty come to MIT today, just like women of my generation came there, believing that civil rights and affirmative action had solved gender discrimination long ago, so that such discrimination would not impact their careers. However, they do believe that the greater demands made on them by family will impact their careers. Today, young women are not willing to choose not to have children, as many women in my generation chose. They fully expect to have children, and our junior faculty do have families. However, these young women, who appear from the outside to be doing it all successfully, told us that they believe that their lifestyle—being a faculty member in science while also being a wife and mother—is so demanding that very few people would choose it. They expect the number of women will not increase unless something changes in the system. We learned, in short, that the playing field is not level for men and women who choose to be professors and have a family.

We learned that as women progress through their careers on the faculty, shortly after tenure they see that their careers are beginning to diverge from those of their male colleagues. The women begin to feel marginalized from important activities and decision-making processes within their departments. Marginalization was the word that came to summarize best the professional experience of the tenured women faculty in science.

These women were extremely successful. If you looked from the outside, you would think, Why is this person complaining? They looked very successful to me. But if you looked at their lives up close, what you saw was that to achieve their success they were working harder and harder as time went by.

They had come in equal, but you could see that as this marginalization happened, the impact of small setbacks accumulated and made their jobs unequal. The men, as they went along, seemed to be doing better and better with less and less effort. They were out starting companies, running departments and labs. Their jobs seemed to be getting easier. What could explain this?

When you looked at the data that went along with this whole process, it was pretty easy to see what was happening. The marginalization that occurred was separating women from important professional activities, intellectual activities, scientific activities, group grants, and administrative decision making, and it was often accompanied by less of other important things as well—salary, space, resources for research.

Now some people ask, How can you make conclusions based on such a small number of women? The answer to this question, as I'm sure everybody here knows, is that it is very simple. The same way as other important decisions are made within universities. Except that in this case, there are a lot more data than are normally brought to bear. Generally, these decisions are made by a man in power, sitting alone in a room, deciding who gets what. Most decisions—about how resources are distributed, what salaries will be, and who gets this or that—are private. Often people don't know what the basis for these decisions is, or even that the decisions are being made. It is a highly confidential process. What the committee did involved a number of people, both women and men, including experienced department heads, looking at a lot of data together. Data had never been collected and looked at in this way, the dean said. Having this collection of data, putting it on the table, and having nine people look at it together, including people highly experienced in administration, made the case very clear.

I want to emphasize how important interviews were. Sometimes people ask, Aren't those only stories? Indeed, when it is just one story it is just that. This is why a lone woman going to her administrator and saying, "This is unfair," can seldom be heard. But when the stories of women who didn't know each other and who worked in different departments, different fields, and under many different administrations came together, the pattern emerged very clearly, the similarity of experiences was startling. It was the coming together of the women in a group that allowed the pattern to emerge and that painted such a compelling picture.

As you can imagine from what I have just said, most of the data had to remain confidential, because the women had spoken to the committee on this condition, and because the number of women is so small that they can usually be identified as individuals from the other types of data collected. The data on space and research volume were considered too confidential by the departments even to be shared with the whole committee and were collected instead within each department, shown to me as chair of the committee, and placed only in the dean's and president's copies of the report. The only data that could easily be released were the data on the relative numbers of men and women at various career levels.

Was there any sign that the number of women faculty was about to change or had changed? Table 9.1 shows the number of female and male faculty in each of the six science departments in 1994, the year we began this work. That these numbers hadn't changed for years is shown by the graph in Figure 9.1, which came to be known as the pancake graph.

I think that often a department hires a junior faculty member, or even two, and they think, We've solved that. Wrong. It would take a lot to impact the numbers shown in this table and graph.

As soon as all the data were collected, we wrote a report and sent it to the dean. As soon as the dean had it, he began to fix things. In the end, he went through and fixed everything that was in the report that could be fixed easily, and he tried to hire more women faculty. There was a great sense that order and fairness had returned. The women went back to their labs and classrooms, far happier now, and almost no one at MIT knew this had happened.

I think had that remained true, then things might gradually have gone back to the way they had been.

TABLE 9.1 Number of Men and Women Faculty in the School of Science at MIT, 1994

Department	Level	Men	Women
Biology			
	Professor	33	3
	Associate professor	5	2
	Assistant professor	4	2
Brain and Cognitive Science			
	Professor	13	4
	Associate professor	5	0
	Assistant professor	1	0
Chemistry			
	Professor	22	2
	Associate professor	2	0
	Assistant professor	6	0
Earth, Atmospheric, and Planetary Sciences			
	Professor	27	3
	Associate professor	5	0
	Assistant professor	3	1
Math			
	Professor	35	0
	Associate professor	4	0
	Assistant professor	14	2
Physics			
	Professor	54	2
	Associate professor	9	0
	Assistant professor	14	2
All Departments			
	Tenured professors	197	15
	All faculty	230	22

NOTE: Data from Lydia Snover, Planning Office at MIT.

But last year, a summary of our findings was published in the faculty newsletter at the request of the chair of the MIT faculty, Lotte Bailyn. This brief article is what came to be known as the MIT report.

We asked the president and the dean if they wanted to write comments to accompany the report. I believe it was these comments, particularly the comment of President Vest, that turned what had been just faculty news into real news. President Vest wrote, "I have always believed that contemporary gender discrimination within universities is part reality and part perception. True, but I now understand that reality is by far the greater part of the balance."

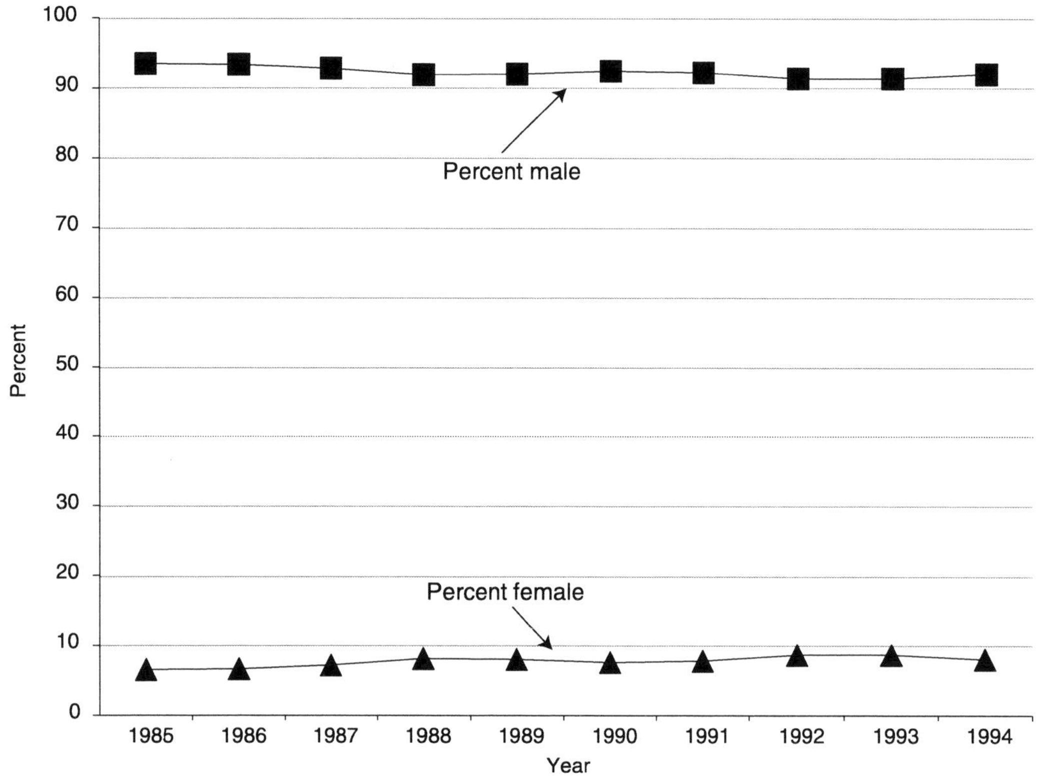

FIGURE 9.1 Faculty in the School of Science at MIT. SOURCE: Data from Lydia Snover, Planning Office at MIT.

I had never thought that in my lifetime I would see the president of one of these universities understand or admit to this reality. Those sentences and the comments of Dean Birgeneau gave the report an impact beyond MIT. A president of one of these elite universities had said, Hey, it's true. That was all that was needed to take a major step forward.

By a small accident, I happened to find myself talking to some journalists one day, and one thing led to another, and I think you know what happened next. Suddenly our story was on the front page of the *Boston Globe* and the *New York Times*. The reaction caught us by surprise and quite overwhelmed us. We were buried in e-mail from women informing us that they had exactly the same problem in their institutions. Many asked us to come and talk to their administrators to try to convince them it was really true. Often women who had been through the same experience had been told by their university that it wasn't true. It was sad. And shocking. The deluge continued for months, and in fact the mail has not stopped a year later.

A couple of weeks after the report was published, I went to the White House, where President and Mrs. Clinton and Labor Secretary Herman said that MIT's handling of this problem should be a model for the nation!

So we found out that our problem was not our problem alone but a problem that is widespread, if not universal, in the workplace. I traveled around this year talking on this topic at other institutions, partly out of curiosity to see what it was like at those other places. When I spoke, I could not tell what fraction of women experience these problems, because I imagine that only women who already understand the issue come to hear the talk. However, among the women I spoke with, the most remarkable thing was the similarity of the experience. When a woman tells you her story, you can finish it for her before she is halfway through, because the stories are all so similar. Women have been telling these stories for many many years—or suffering in silence with them.

Where do we go from here? Let me begin with MIT. I am very optimistic about the situation at MIT because I think at MIT two of the most critical components for success have come together: First, the president and the provost are completely supportive of the women faculty and want to make significant progress in this area, and they have come to have a very sophisticated understanding of the problems. Furthermore, they realize that it is essential to work *with* the women faculty to solve this—not solve it *for* them without the women faculty's participation, as often happens when men say they are going to solve the problem. Second, the women faculty at MIT—not just in Science but also in Engineering and in the other three schools as well—are remarkably cohesive and involved. I think this is almost as crucial as the administration's support. In my travels I learned how hard it is to achieve this. Young women often don't want to be involved in this type of initiative because they want a chance to make it on their own. We must give them that chance, and in fact I think we should avoid involving young women unless they seek us out. Often older women faculty are afraid to be involved in this sort of activity for fear of damaging their careers, and one cannot blame them. Then there are women who truly do not have the problem and do not feel an obligation to help those who do or who do not understand the matter any better than many men might. For some reason, the women at MIT, regardless of which group they fell into, did feel an obligation to take this problem on—not just for themselves, but for others. I can't explain what happened to produce this result, but I am very grateful for it.

So at MIT now there are equity committees in all five schools, and they are collecting data and so on. I think this is a terrific thing. At least we can ensure equity in measurable resources and rewards. The only problem we have is those 800 faculty who have not been involved in these activities at all, at least some of whom either don't know this happened or don't believe it is true. Somehow we have to raise awareness about unintentional gender bias, and that of course is the real challenge. However, I think that, with institutional support and change, we have a real chance at MIT. But the job is hard, and the question is, Can we do more—both at MIT and perhaps also outside MIT? What is a more general solution to this problem? I don't know, but I would make the following comments.

Women have known for a long time that two major reasons for their underrepresentation on science faculties in the United States are gender bias and the greater family responsibilities that fall to women. Many studies like the one at MIT have documented gender bias, and we have heard at this meeting that it is a well-known phenomenon in many fields beside science, and that study after study confirms this. This is not about affirmative action, this is about discrimination and exclusion in a profession that requires extensive interaction.

At this meeting we were told by a representative of the highest level of our government that the economic future of the United States, to say nothing of the prosperity of large segments of our population, depends on increasing the number of women and minorities in science and engineering. This seems so odd, given the fact we have spent 2 days talking about how women are being driven out! Given these contradictions, it would seem to me that the following should be done to address gender bias:

- We must call a moratorium on further studies that try to document gender bias. While gender bias is not universal, we must accept that it is so common a problem that we must institutionalize the actions needed to guard against it, correct its consequences, and raise consciousness to ultimately eliminate it—in both men and women.
- Institutions must accept responsibility for eliminating gender bias. Now that the problem is so

well known, the leaders of our universities, government agencies, and academies must accept the responsibility to fix it.

- We must decide that men and women will share decision-making roles at every level of our institutions, and that women and men together, chosen by the women faculty as a group, will continuously review data pertaining to resources and compensations to ensure equity.
- Women, and men, who perform this work for their institutions must be compensated just as any other administrator. Otherwise, the message will be clear that this work is not really important, that the institution does not really wish to fix the problem.

To address the fact that greater family responsibilities fall to women, we must recognize that most high-level jobs in science were designed for a man with a full-time wife at home, a situation that does not apply to many men and most women today. We must restructure the job and provide support, so that both men and women can work on a level playing field. How do we do this? Again, I do not have the answers, only some suggestions. First, I believe that this is a national problem and should be taken on at that level, not just for women in science but for women and men in every profession—not to mention children. When so many women entered the workforce, we forgot to ask who was replacing them in the home. I suspect that it would cost about $150,000 a year to replace a college graduate in the home—and that is probably an underestimate. Of course there have been many adjustments made in workplaces, but I believe not enough. In particular, what might we do for science?

- We should insist that day care be built into every laboratory building.
- We should make child care a line item on grants.
- We should revise the grant system so that people do not spend such enormous amounts of time raising money.
- We should continue to review the structure of our academic jobs to fit with family needs and with our longer working life. Many people expect to work for 55 years now. Children may need our greatest attention for only a small fraction of this time. How do we structure the job to fit the way life really is and to allow *both* men and women an equal opportunity to have families—*and* to win the Nobel Prize.

Finally, I would like to say that I believe we are moving far too slowly in fixing these problems—both for women and for minorities. I hope this meeting—and so many others like it—will serve as a call to action for those in leadership positions in all fields of science and engineering in this country.

I don't think the underrepresentation of women in science is such a mystery. I think the question is not so much why women leave science—we know many of the reasons why. I think the question is, Now that we know, do the people in power have the will to fix the problem?

DISCUSSION

Linda B. McGown, Duke University: That was a wonderful story, and you have done a very remarkable thing. Thank you. I am trying to figure out exactly what stage of denial I'm in. I'll be thinking of that on the plane home.

I think one of the important points is that secrecy is really the enemy, particularly at a private institution where nothing is published. I find secrecy to be a constantly moving target. If I complain about my salary, I can be told anything, and all I'm given are averages. We all know that averages don't necessarily represent the actual distribution in terms of who is up here, who is down there, and how many are in between. It tells only a very small part of the story.

I have argued in the past that I thought there were salary inequities, but I was only given summaries of data that did not address the problem. Space is pretty clear—we have a small department and I can go measure it myself. Our chair recognizes that this (space distribution) generally has been public information.

But everybody cuts deals with the administration. I have to provide my own matching funds, which defeats the purpose of having overhead recovery, and I'm sure other people have the same stories.

The problem is that in a private institution, none of this is a matter of record. If there were some way for federal agencies to compel this information to be revealed, then people would have to behave just a little bit better or face the consequences of these sorts of appalling inequities. Something needs to be done, and I think publication of this information should be mandatory. I don't care if people know my salary. I think it is vital that this information be out there.

Nancy Hopkins: In principle, I agree with you, and I used to believe exactly as you do. Having gone through this process for 5 years at a place like MIT, with its secrecy aspects or the privacy or whatever, I'm not sure that I wouldn't rather have women with a completely diverse power structure, where people of all types are integrated throughout the system. That would be almost as effective, if not more so.

In other words, women really rock the boat. That's tough. Just now, a woman has been in the actual primary salary data in the School of Engineering for the first time in the history of the school. To me this is a major breakthrough. It didn't rock the boat; MIT is still standing. I think that it can be done within the same system if it is a truly integrated process like that.

This is an alternative way. We learned an amazing thing—that even in schools where the salaries are published, women are still underpaid. Even in cases where they have sued and won. You can't believe it, you just can't believe it.

Suzanne E. Franks, Kansas State University: One of the things that has been an ongoing theme, certainly of this workshop, is what I will call the liberal project in science, which is "increase the numbers." We need more women—increase the numbers.

There has been some discussion about changing the structure and changing the environment, and I think certainly your presentation has touched on that. I will call that the radical transformation project, which is not just to increase numbers but to change the structure and change the kind of people—change ourselves in the way that we look at science and the way that we look at our dedication to science.

As I work in my position to try to encourage young women to stay with science and math and to major in engineering and science and go on through the pipeline, one of my concerns is, What can I do, not just to increase the numbers of women who do this kind of work, but to have an effect on their attitudes?

I think we need not just more women scientists but more feminist scientists. I wonder if you have any thoughts on what the people at this workshop might do to make it possible for women to be feminist scientists, and for men as well.

Nancy Hopkins: I think you probably have to ask Debra, or other people. I think I'm the wrong person to answer that question. I hear you, but this is why I think we were so surprised by the MIT study. I think the people who went there all bought into the system. That is what is so strange about what happened there. I think it allowed the gender bias part to stand out. These were people who didn't have family issues. They had bought into the system, they weren't trying to change the system. They were just trying to be part of the system. So it allowed one to see the problem of discrimination separate from these other issues.

I think what you are talking about is something we never discussed. For Lottie Baylon, this is her professional work. It involves changing the structure of the workplace to make family and work much more compatible. We have argued over this for years. It is a complex issue. Something is going to have to be done. It is going to take young people saying how it has to be done.

Christine S. Grant, North Carolina State University: Last night we spoke extensively about this. I want to thank you and the women at MIT for doing this study. I am at North Carolina State, where Marye Anne Fox is our chancellor, and the two of us have spoken briefly about trying to do a similar study at North Carolina State; there are probably other women in the room thinking they should do it at their schools, too.

One of the things that I wanted to know is if there is a guide—not that you need more work. I am sitting here at University X, I want to do this at my university. After 10 years I recently did my space list and said, "This isn't good." It was 5 years ago that some women grad students came up to me and said, "This isn't good." I don't know if I was in denial or just trying to get tenure or whatever. But now I'm looking at it saying, "This isn't good."

One of the things that we have talked about is having a consultant come in and do a salary equity study. Marye Anne Fox mentioned that possibility at a recent meeting of women faculty from the whole North Carolina State university campus. We are trying to form a women's faculty organization, and we had a luncheon a few weeks ago.

So I have two questions. One, Is it possible to get this, and if so, how? Two, What do you think about hiring a consultant rather than having equity committees within thc university?

Then one last comment about the previous comment by Suzanne Franks. The women faculty at North Carolina State have struggled with this issue of putting the word feminist in the mission statement. We could probably stay here and talk all day about this. The women scientists, myself included, were saying no, because we are going to do this anyway. Maybe it could be in a statement about what we're trying to do, but we shouldn't have it in the mission statement, because it might scare off women.

Nancy Hopkins: On the question of how to do the study, you can see why it was never published and never will be published, because it is obviously very transparent—you can identify all the people and everything about them. So that can't be made public. We have talked a lot about writing something to describe how we carried out the study, but we just haven't had time. We are going ahead with a new initiative, to make this whole thing a more formalized structure, an alliance between the women and the provost and the president. If we do that, and if we raise some money to help us, we could actually write something about the process—because people have asked for it.

I don't feel that we are experts. I do think that each institution has to guide its individual process, adapting and designing it. In a way, the process of designing it addresses the problem, because you are talking about it and engaging the administration.

The key is whether the administration really supports it or not. A lot of people ask, What do you do if they don't? And I just don't know what you would do. In our case, it was really hard in the beginning, because we did have to apply quite a lot of pressure to get it started. So it is a tough problem.

As for getting outside people, no, I don't think you can do it. I think these big statistical studies just bury what you are trying to find. In our study we had a physical chemist look at physical chemistry space and a biochemist look at biochemistry space. You really have to understand the field to know why it matters. If somebody's cauldron is in the basement and they are extracting up on the fifth floor, you have to know a lot to understand why this arrangement could be so damaging to somebody. How could an outside person come in and do that?

That is the problem, and it shows why the process is very time consuming. It must be recognized as serious administrative work that is compensated by time off from teaching or time off from other things. It is not trivial, casual work. It is time consuming and hard.

Sally Chapman, Barnard College: I am one of the people who read the documents—that is, what was available on the Web—with tremendous admiration and excitement.

One of my reactions then is one of the questions I still have now. I was stunned by MIT's admission of this. You read these little snippets from Stanford and Harvard and so on—denial, denial, denial, institutional rather than personal denial. In this era when people are terrified of litigation, I was stunned that MIT was willing to step up and say, "We made these mistakes," and that some lawyer didn't say, "You can't do that, because if you do that, you're going to have all kinds of lawsuits, and you're going to have to pay out vast amounts of money." Did this become part of the discussion at any point?

Nancy Hopkins: I asked the president why he did it, and he gave a really simple answer. He said, "It was the right thing to do, so we did it." He said, "We didn't ask the lawyers." All the other university presidents asked, How did you do it? Didn't your lawyers tell you not to do it? That is a sad comment on the whole state of the legal profession, that the people who are supposed to be helping us are preventing this problem from being solved. This is really sad.

It has turned out there were some inquiries right after it came out from people who had worked at MIT in the past and called up and asked, Can I use this to sue MIT? I said I didn't know. But there hasn't been any serious effect from it. It is a very sad thing. It is incredible. It is the thing I find most amazing, when there is this ongoing investigation.

You would have thought the Frances Connolly book, about walking out on the boys, that alone would have caused them to say maybe they had a problem. It is incredible. Now people are saying the future of the United States depends upon these underrepresented groups. What is going on? There is a disconnect. I just don't get it, do you?

Maryka H. Bhattacharyya, Argonne National Laboratory: I think when we have a problem and are looking for a solution, it really helps to narrow down what that problem is. I just wanted to make the analogy that I made for myself and see if it makes sense for anybody here. That is, in 1850, when the women at Seneca Falls were trying to get the right to vote, all the power was in the hands of males, in terms of the decision that was going to be made, but women were half of the population. We only had the power of persuasion, that it was the right thing to do, and that all of society would be better off if both males and females had the right to inherit wealth, the right to education, and the right to vote. It was by the power of persuasion that we could get that vote, and it took 70 more years to get it. If you look at the history of that effort, women had all different perspectives. There were women who said, "I'm going to lose everything if I fight for that cause. Where we are is better." There were women who said, "It has to change. How can we leave it this way?" It was an up-and-down effort that progressed slowly over time.

When I first came to Argonne and was working for the Women in Science and Technology (WIST) program there, the progression that you described was very similar to mine, with that same optimism. We were really going to change things. We got funding such that for a 2-year period, women in R&D could work for the WIST program and be paid 30 percent of their salary to do that, because there were so many undone things. We got that money from the Office of the Director. We went to the top of the organization to get it. We felt it was going to make a difference.

Now at Argonne we are still 10 percent female on the staff. In a 10-year period, nothing has changed. Over a 40-year period, it has not changed. It started out at 8 percent, and maybe now it is 12, but it is in that 10 percent range.

Using the same analogy I made about getting the right to vote, we are only 10 percent but we have the power of persuasion and the power to change things in a male-dominated organization. The importance of persisting is clear. Because of the nature of the problem, it will not be solved quickly. And yet, the strategies that you discussed to show how things can change in one place are inspiring. We need to persist over time to change a system where we are in a small minority.

Nancy Hopkins: May I ask you a question? There is something that really concerns me a lot, because we all talk about numbers and we all know that we would like them to get larger. And if we could get them larger we would have solved the problem.

By definition, minorities will always be "less," that is what a minority is. What bothers me is our inability to treat people and integrate them, regardless of how many there are. I don't care if there are 10 percent women; if that is all that want to do it, if those women are incredibly happy and having a great time—I am of two minds about this.

Maryka Bhattacharyya: I was more or less pointing out what a difficult job it might be as 10 percent of the whole to bring that persuasive argument, because seeing that difficulty can help you to solve the problem. But I like your analogy a lot, your saying we will always be a minority, and part of the problem is to solve the issue of diversity in the workforce and to treat everyone well.

Robert S. Marianelli, Office of Science and Technology Policy: Since I got involved in this workforce report,[2] I have learned that the minority has become the majority in these 50 years in the United States. I think people sometimes underestimate the power that they do have. After all, everybody pays taxes and votes. So if you are aware of what is going on and it doesn't make you very happy, you can do something. I doubt that people will continue to support things that they aren't a part of as the demographics of the United States changes.

It's just a thought. Maybe there is a way to use information to engage the right people to force some changes.

Margaret V. Merritt, Wellesley College: In reading the reports that appeared on the Web from your study [at MIT], I was impressed with the very careful data analysis. In fact, the report was somewhat dispassionate. I think what was really wonderful today is to see some of the passion behind those data.

Nancy Hopkins: You ain't seen the half of it!

Margaret Merritt: One of the things I would like to follow up on here is your remark that one cannot be a scientist and be marginalized. I think that probably is at the heart of the matter here, that your science really suffered during that particular time. I think those of us who have been in circumstances in which we felt marginalized understand the reason for your anger: that your career, your science, suffered under those conditions.

I think that this marginalization remains the important aspect. We must deal with it—regardless of

[2]Commission on the Advancement of Women and Minorities in Science, Engineering, and Technology Development, "Land of Plenty: Diversity As America's Competitive Edge in Science, Engineering and Technology," July 13, 2000. Available online at <http://www.nsf.gov/od/cawmset/>.

whether it affects 50 percent, 20 percent, or 10 percent—because it limits the quality of the work that women scientists can do.

Nancy Hopkins: Exactly. I think what I learned was how small each decision is, but how fast they accumulate to have a serious impact. It goes back to the idea you had that you could succeed if you were really good enough, even if you didn't have a lab. You would be out in the hall doing your experiment. But it isn't true.

Margaret Merritt: If it is a slight inconvenience, I can do it.

Nancy Hopkins: Not at the top. Not when you're in an entirely competitive, fast-moving situation—of course not. How silly it was for us to think so.

Marion C. Thurnauer, Argonne National Labs: I would just like to comment on your description of working within the system to effect change. Maryka Bhattacharyya mentioned our WIST program. Ten years ago, when we first approached laboratory management, we had an experience similar to the one you described. We also felt we were walking into a room full of intimidation. But after we had worked for over a year with the laboratory management, I came to realize that laboratory management was just as apprehensive of our initial encounters as we were.

I want to make this point because over the course of that year we developed a mutual trust. I think this is a very important thing that has to happen before you can effect a change. I think you are an example to all of us, and it certainly helps us at Argonne to see that an institution like MIT has done this. I think it will move us another step forward. The Argonne program was launched 10 years ago, and we have kept it alive, but we need now to move forward in many ways. The existence of a program like yours is also very important for others. So thank you.

Nancy Hopkins: Thank *you!*

Rosemarie Szostak, Department of the Army:. When my career started in the early 1980s, I had two job interviews (it was a bad time to look for a job), at MIT and Georgia Tech. MIT, bad environment, so I took Georgia Tech. Bad choice.

The bottom line (because the Army teaches you to get to the bottom line first and then get to the details) is that transparency is what is missing. What we don't see and what the departments need to do is to explain what the rules are.

There is no secret, especially in a state university. That was one of the things I found out with my lawsuit: if there are no written rules, the guys can make them up as they go along.

So I have done the exact same thing you did. I did a detailed analysis. I published it in the university newspaper. Nothing happened, except that women stopped talking to me after that. But I did find out what actually happened. The administration went to the attorneys and said, "We have to deal with it." So it went badly from there.

But the point is, you need to push transparency in the hiring practices, in the way promotions are done, and in the way labs are assigned, including square footage. If I knew that my square footage would correspond exactly to the dollars I bring in, and if I knew that everyone would be in the same boat, that is equity number one, and I would be a lot happier.

I think the problem we are seeing in the universities is the secrecy and the different rules. If the rules were published and not secret, we would see a lot more equity.

Barbara J.W. Cole, University of Maine: Listening to the comments of the last few days, it struck me that I have been in academia for 14 years—my whole career, right out of graduate school—and why on earth am I here? There has been a lot of very negative talk about the academic life, particularly for a woman faculty member.

So I guess I wanted to say that I really like my job. There are quite a few untenured faculty members in this audience. There are good places, there are good jobs. I think men go through times when their jobs stink and so do women; there are cycles that we all go through. But all in all, an academic life, in my own personal experience, has been great.

I am currently the chair of the chemistry department at Maine. We have three women on the faculty. Two of us have children; one is in her first year and is pregnant and due in August. I had cribs set up in my office and nursed my children in my office for the first 4 or 5 months, until they could get into day care. I had a lot of support from male colleagues, and those that didn't think too highly of it didn't have the courage to say so. So I think we don't always have to look just to women for that support.

But I also found that if I didn't ask for permission, I also got away with a lot more. I didn't ask the chair of the department what he thought about me setting up a crib in my office and taking my son to faculty meetings. I never once asked, What do you think? Or, Would that be okay? And fortunately he didn't have a problem with it, or he never said anything.

So I think part of it is, we shouldn't ask so much. We should just do it. I would encourage particularly the young faculty members to do that. The powers that be are going to have a lot more trouble saying something after the fact. You put the burden on them to make an issue of it instead of making it an issue yourself first.

Nancy Hopkins: This reminds me of one great story at MIT where the same approach was taken. A man wanted some space and they didn't give it to him, so he just took a sledgehammer and took the wall down. They thought it was great. They loved it.

Barbara Cole: I like that. Also, I am in a very lucky situation, in that the women are more highly paid than the men at my institution, although it is quite dependent on the specific department. But in our department we do quite well.

I think the biggest problem we—all women—have is because we're trying to balance everything. We probably have slightly lower publication rates. The agencies like NSF, NIH, EPA, and DOE come back and say, "You're at Maine, you have to teach." I teach general chemistry to hundreds of students every year, as well as graduate courses, and I have a big research program. I think there is a perception at the agencies that people in those institutions aren't very serious about their science.

I find it to be quite the contrary; we are serious. We don't spend 20 hours a day exclusively doing research. We do a lot of things. I think that some of the bigger, more prestigious institutions would actually have much healthier people—and probably more creative science occurring—if they would allow some of that.

Cecily C. Selby, Radcliffe Institute for Advanced Study, Harvard University: I think I'm being allowed a postscript to remind the audience that the science community does have a history of turning against even the most frustrated scientists who communicate well to the public. We must not let this happen to Nancy Hopkins. I know her work is absolutely top of the line. I think we should continue to monitor her science as exquisitely as it deserves, while recognizing also what she is doing for her colleagues.

10

Reports from the Breakout Sessions

Following the presentation described in Chapter 9, breakout sessions were organized to enable more extensive discussions among the workshop participants. The following questions and statements were suggested to the breakout groups as possible topics for discussion.

- What are the unwritten agendas and folktales of career prospects for women in chemistry and chemical engineering? Are they realistic?
- Are there rewards for an institution that improves its record with respect to hiring and promoting female chemists and chemical engineers?
- How rapidly can organizations change? How rapidly must they change?

Discussion leaders from the breakout groups then reported in plenary session what they believed to be important ideas and topics that had emerged in the discussions.

Frankie K. Wood-Black, Phillips Petroleum: Our group began by talking about some of the folklore issues, and then we jumped to the last of the proposed questions. Do you remember the last two questions? How rapidly can this change happen, and when should this change happen? Yesterday would not have been too soon. But the reality is, how quickly it can happen will depend on the structure that you are dealing with and what the drivers are for your particular organization. Nothing will happen without the drivers.

We then talked about some of the different agendas and folk tales that are out there. Where are the students going? This looks like an issue that several of us are going to take back to the various groups that we work with.

A common piece of folklore is that students receiving M.S. degrees are now available to go to industry, and that industry is coming in and hiring at a progressively higher rate. There are computer jobs; we all know the dot-coms—dot-coms are taking everybody. I hear anecdotally from parents who want their kids to get a 4-year degree, that if the kids are good in computer science, they get an internship during the summer and forget about going back to school because they can make a million dollars in 2 years. There is a woman chemist in Tulsa who says she was able to organize a start-up because her son had already established a dot-com, and he was able to sponsor her start-up.

Another anecdote concerns whether students receiving B.S. degrees are going into medical school at a greater rate. That question is there, and we need to look at it. Data will show the reality, but as you have seen over the last day and a half, the reality isn't so hot either.

We spent most of our time talking about rewards and where they were. One suggestion was to generate a list of how different departments rank with respect to women. If such a list were published, would that really work and would it drive what you were after? I'm not so sure. Sometimes the negative press would dominate, and sometimes the positive press would.

We talked about the structure for research grants. The big driver is the pocketbook, so where is the money coming from? Is it coming from the federal agencies, is it coming from the state organizations? And there is accreditation. We talked about the ACS Committee on Professional Training (CPT) and whether or not there should be a question or a focus issue about women in the accreditation process.

It came up continuously that everybody has become so politically correct that you worry about the legal whiplash. If we do some things, we may be opening ourselves up for a counter discrimination suit. So possible legal issues may create a barrier as we talk about things—we need to either stand to the side or take the risk. And if we let the lawyers at it, we all know what the lawyers are going to do.

Awards are another issue. We focused on big awards, because the small awards are merely a bit of recognition that could be funded out of pocket. What are these rewards supposed to drive? Presumably, their goal is to change the infrastructure.

We probably have the equivalent of the old route system. Remember Route 66 and Route 25 and Route 50? Frequently, that wasn't efficient to get where we needed to go. What changed? We developed the interstate highway system, which took a massive change in thinking. Now we are trying to change the infrastructure to help women move forward.

My final point is, "Do something." I'm going to leave you with a piece of evidence for which we actually have some anecdotal support. It is okay to fail. It is okay to try and fail and say, "This didn't work for this situation, and this is why." It is not okay to not try. You must go out and try. Just sitting and doing your own thing will not be a catalyst for change—it will not be an agent for change and nothing will happen. We all must do a part. I can tell you from my position as chair of the ACS Women Chemists Committee, that we have been trying to get women recognized with ACS awards. We have been following the numbers, and we have been trying to get women recognized. But I can tell you, it doesn't matter how much preaching gets done, women are not participating in the ACS awards process. The numbers are appalling, when you start looking at how few women submit a nomination document. This is a problem that we all can work to fix without having to get involved in a group, without having to spend a whole lot of time. Sometimes we only talk; we need to be out there making the change ourselves.

Cecily C. Celby, Radcliffe Institute for Advanced Study, Harvard University: We also focused on organizational change and asked why we were interested in it. We have been talking a lot about symptoms of climates that are unhealthy for women, so why are we moving from symptoms and disease to organizational change? It is because we talk about all the sad tales—but productive talk about sad tales, as the symptoms of an underlying pathology—and to meet our goals, we have to look at institutions as a whole.

As we look at institutions, we notice that one of the symptoms of disease is failure, although as a biologist, I'd rather talk about failure to adapt. We all know that the healthy surviving organism is the one that adapts to changes in the environment. So looking at institutions, there is a very simple conclusion—they only change when the leadership is effective and strong.

Focusing for the moment on academe, retention of talent is a major question. We know there are problems with retention of talent at all levels, and all kinds of talent—not just female talent. What do we know about the retention of female talent? I think the reason for dropping out can be traced to when the climate in the workplace is discongruent with the person's values—when the behaviors in the workplace are discongruent with the person's own behavior—women more than men will opt out. So I go on that assumption, and we have that kind of data.

Looking at institutions as a whole, one question is, Can they change rapidly? We had one or two examples, both academic and industrial, of institutions changing very rapidly when the leadership wanted it to—when the leadership was committed to the kind of goals we're talking about.

How soon? I think our message has got to be, the sooner the better, not just for us, but for science, for technology, for men and for women. What is good for women is good for science and engineering and good for men.

How will we know what we will want to change? What are we after? How will we know that we have succeeded? It was mentioned that one of the signs of success will be when some of the goals suggested in this workshop can be seen at an institution—when individuals can see themselves represented in the institution, either as students, as faculty, or as administration. I can even add corporate board members. In other words, we'll know we have succeeded when everyone who wanted to succeed, succeeds. This is not to say that everyone should study science or everyone should want be a scientist. But for those who have made the choice, they will have the opportunity to succeed—the climate is favorable.

We can even go beyond this and say that our real concern is, Who will do science in the 21st century? This is important. My particular point of view is that we need to get the public—in the broad sense, from taxpayers all the way up to leadership—to care about who does science. This has to be done for the sake of science and for the sake of society. Who is going to choose to do science, who is going to be enabled to be successful? If they are going to be successful, they must have the kinds of environments that are congruent with their values, their ambitions, their tastes, their styles, and so on.

We discussed diversity quotients as a strategy, talking about the value and limitations of quantitative data, and about the transparency of data. That is a very important issue, as we all know today, because of the arguments about privacy versus transparency. In the best cases, private companies are required by law, by practice, and by stockholders to be much more transparent than academic institutions have to be at the present time. To what degree will social changes be forcing more transparency on academic institutions, and what are the implications of that? Take MIT's "secret data." What implications would such transparency have for faculty privacy? There are many privacy issues.

A major concern is diversity quotients. We could use them in a variety of ways. But how could they be constructed to include not just numbers? How can we build in attitudes, values, perspectives? Could we rank institutions like the automobile industry ranks automobiles? They find a way to rank automobiles by a whole variety of criteria, some quantitative, such as drivability, and some aesthetic. And we evaluate our students by quantitative and qualitative data, so we ought to be able to do that.

These last points respond to what was suggested this morning by Ms. Sendall, who mentioned the business reasons to motivate needed or desired change.[1] I've suggested some academic reasons to

[1]A written contribution for this presentation, "Gender Diversity in the Workplace: The Leadership and Organizational Imperative," was not available for inclusion in this report of the workshop.

motivate academic change. However, there are complications that I don't believe we could possibly have talked about in detail in this meeting. Those who are ultimately in charge of academic change are the boards of trustees, and those who implement academic change are the deans and faculty. But the beneficiaries of academic change are students, so we have a very complicated constituency. I remember years ago learning that one of the troubles in academe is that you tend to have very, very conservative board members and very liberal faculties. In other words, the interests of the board and the interests of the faculty are not necessarily congruent.

My personal plea is that you find academic reasons for change that will motivate boards, deans, and faculties and be understood and valued by the students. I suggest that we must add to our current arguments about human rights, equal opportunity, and human capital the argument that everything we are talking about will be good for science and technology in the future—and certainly for the future scientists and technologists. I would love to see us think more about how diversity could help our fields.

Barbara K. Warren, Union Carbide Corporation: I would add that we talked about how to increase the representation of women at intermediate levels in academia and industry. My suggestion is that we might share examples, and we might actually encourage management in both universities and industry to take chances and bypass some of the bureaucratic procedures, some of which may hurt women. A lot of exceptions were made for men in the past. Since we see a big need to put women at intermediate levels, why don't businesses and universities take the chance, give us the rope, and allow us to risk hanging ourselves? Why don't they go ahead and make exceptions and see if we can do these jobs? They need to think creatively about how to promote women. Give women the chance to fail in these intermediate positions. Try to make progress and do it quickly.

W. Sue Shafer, University of California, San Francisco: We looked at a number of myths. I think one of the myths already mentioned here is that the civil rights movement and the EEO movement have solved all our problems. I think we have heard enough to realize that is not the case.

There was one myth I hadn't heard before, but which could be insidious. One young woman was told by a male colleague that one of the dangers of women going into academia was that they would end up being divorced. It appears that this myth is being used to discourage women. I guess it shows how a comment made jokingly during a cocktail hour conversation—"If you come work for us, you're probably going to get a divorce" —can have a profoundly negative effect on someone.

Another myth is that women who have children will inevitably fall behind. You have to decide for yourself if you want kids, and when you want them; it's a very personal decision. We know people who have succeeded and you hardly realize that they took the time to have kids. But on the other hand, we have also heard about people having children and having difficulty getting back into the laboratory, getting proposals funded, and getting restarted. The Office of Research on Women's Health has programs for people who need to reenter the workplace, for whatever reason. But they are programs that have fairly specific rules, so you might want to talk to people at NIH about them, if that is an issue for you.

Another myth is that women have research grants but get fewer dollars per research grant. When I first went to NIH, that was true; however, when I left it wasn't true. The reason it was once true is that women used to ask for less. They have gotten smarter. So at least on the individual grant basis, they are no longer funded less frequently (in terms of success rate) or less generously than men.

The question was raised in our group about whether women have fewer grants than men. I'm not certain if they do or if they have a harder time than men getting the second or third grant. I hope we look into that.

We have heard it asked if industry can use the bottom line as a justification for increasing the number of women. If the academic sector is more intellectual, does this mean it ought to increase the number of women because it is the right thing to do? I think the answer in both cases is yes, but you can't necessarily expect academics to be more altruistic than the rest of the world.

One member of our breakout group described a program funded by the Luce Foundation, which evidently targeted institutions that had poor records of having women—I'm not sure if it was in chemistry or science. They made essentially a 5-year chair award that would give summer support for scientists and support for other things such as undergraduate fellowships in science, women in science offices, and so on. This has the effect of asking an institution to make a commitment to increasing the number of women in science. Is that the kind of program that could be extended further and that could be considered by agencies such as NIH or NSF?

Another topic that came up was about scoring departments on their diversity standards and somehow connecting their eligibility to receive more funding with these improvements in diversity standards. For instance, if there were a report card on diversity standards that could be used by universities, might this be a better tool than federal enforcement through EEO or civil rights legislation? If you give an institution a positive incentive instead of bringing in a big federal bureaucracy to try to do this, the results might be better. Nevertheless, there is much to be said for the notion of possibly using Title IX as a weapon.

It was noted that the National Science Foundation has some requirements—which many of you know better than I do, I'm sure—for describing a department's atmosphere and diversity. Many times, these requirements are not addressed in the applications, but they are supposed to be addressed in both the applications and the reviews. Perhaps NSF could be stricter about not reviewing applications that don't appropriately describe such things and perhaps it could give reviewers more direction as to how they should be weighted in review.

There is anecdotal evidence that search committees tend to have their own agendas and that they nitpick in an effort to be seen as evaluating things appropriately. Sometimes this can reach ridiculous levels, not only discouraging applicants but also ending up with the wrong people. So one needs to look at how one is running one's search committee and how even-handedly one is treating men, women, and minorities. When somebody does something outrageous, you need to call them on it right then and there.

Another way to effect change is to organize brown-bag lunches, to get the word out on things like the discussions at this meeting, and to suggest actions that people could take. One member of our breakout group reported that she had volunteered to organize her departmental administrative review, because that was the way that she could get her agenda items considered along with those of the rest of the department.

Finally, "Network, network, network." Sound familiar? We talked about enlisting males and people with star power as allies in what you need to accomplish.

Are there ways in which we could reward departments that produced good students who go on to take positions in academic departments? I didn't hear any specific suggestions on how this could be done, but it is something we all need to think about.

A member of our group said that one of the most important things in his development as a scientist had been the opportunity to go to regional meetings for students and faculty, where he and other graduate students could start early to present their research in a friendly environment. That might be something to consider organizing for women faculty and students, as a way of reaching women particularly.

To summarize, I think there are two areas where we need some more data. First, does having a family delay career progression, and if so, how systematically? And second, are women as successful as men in getting a second or third grant?

I think we need to *do something*. We each need to take home three things that we think are the most important things we can do. They may be local, they may be regional, they may be national. But I urge you on your flight or your drive home to think about the three most important things you could do. Write them down and carry them around with you until you have checked them off. Then I think you will have accomplished something.

Maria K. Burka, National Science Foundation: We are, as you may recall, the group that has chemical engineers. One of the first comments we heard was that there aren't any folk tales in chemical engineering—because we are the first generation of women in academia. But as the discussion went on, we discovered there were some folk tales. We started by discussing the notion that to be treated the same as men, women have to be two and a half times as good. It isn't clear whether that is reality or myth.

Another folk tale is that you should have children only after you have tenure, because once you have tenure, life becomes easy and things are much smoother—you don't have anything to worry about. That certainly does not seem to be the case.

It was reported that in some schools, these problems don't exist anymore because there are very enlightened people in leadership positions—deans, department chairs, and so on—who have taken the proper steps and are addressing the situation. We should be fair and give credit where credit is due. The problems are not universal. There might be problems in some departments, but other departments are really taking the proper steps.

When you have one woman in a department—which is very often the case in a chemical engineering department, and very often in industry as well—that woman is appointed as the token woman on every single committee. This takes time away from her scholarship, it takes time away from her research, and ultimately—when she comes up for tenure or promotion—the response is, "Well, she hasn't performed." But the woman's time had been sapped by the activities of all these committees to which she was appointed. In some schools, department chairs have addressed this problem and put a stop to the practice. Nobody in that department can be put on a committee unless the department chair agrees, and the department chair will shield women and minorities, so that they are not penalized by having to sit on every single committee.

Many students believe that because you are a woman, it is easy to get a job and it is easy to get promoted. This is a serious problem that we do have to address, because the misconception exists on a lot of campuses.

Some people argued that it is not fair to paint all men as villains, because there are many men who just don't understand what the problems are and who are looking for information. The reality is that if we communicate, we can handle some of these problems without a confrontation. What about the idea that men over 50 are the enemy? Again, it was argued that this is not accurate, because there are many men over 50 who are in fact very anxious to mentor and help. It is really important to keep in mind that the pull at this time has to come from white males, so we all have to work together.

Women have a societal problem that is not confined to chemistry or chemical engineering or to universities. Many women have low self-esteem, and a lot of them feel that no matter what they do, they are wrong. In other words, if they have a career and they have children, they are not doing a good job of either. On the other hand, if they stay home and raise their children, they are asked, Is that all you're doing? This is a big problem, although I'm not sure how we can solve it.

There is another folk tale—that fixing the pipeline will do the trick. This is something that is constantly used against women and it is not fair. We should not permit that to be used, because the pipeline problem is not a valid excuse.

Another point that was mentioned was the feeding frenzy when you are out there looking for a job. Everybody wants you. The companies and the schools make these great promises. Then as soon as you're hired and you've signed on the bottom line, the honeymoon is over. You can't get the lab space, you can't get any of the things that were promised to you initially. This is a big problem, particularly for women, because men seem to be better at camping out and demanding that they get the lab space, demanding that they get the promotions. Perhaps women are not trained to be quite as assertive.

Our group discussed rewards to the institutions to improve their records. For industry, the ranking system is important, because executives pay attention. The cover of a recent copy of *Woman Engineer* magazine showed rankings for the top 50 companies—how welcoming they were and what kinds of places they were for women to work at. Companies paid attention to that.

We need to talk about work quality issues. In industry, managers are compensated according to how good a job they do. This is not necessarily the case in academia, and this might be something we could learn from.

We also talked about the trickle-down effect. That is, if companies perceive that a university is not as diverse as they would like, they don't go there to hire and they don't give donations. So academic institutions do need to keep these things in mind, because this is the reality.

It was suggested that the reward system at academic institutions is much more tenuous. Louisiana State University has just graduated 12 African-American women with Ph.D.s, and this has made companies take notice. They are sending interviewers there, because there are people they want to hire. So there is a positive pull.

Then there is of course the pull that students are needed for the future, in the face of a decreasing number of applicants to graduate school. Universities do have to take notice, because if they can't get the graduate students they need, they are going to have to be more diverse.

We talked about the *U.S. News and World Report*, which is the bible for a lot of academic institutions. They wave it around, saying, "We're number one." I know whenever I visit a school, it is astonishing: as soon as that issue comes out, the first thing every department chair says to me is, "We're number two" or "We're number four." There is a real possibility that if the ranking included diversity as part of the metric, people would sit up and take notice.

We also discussed something negative—the backlash. When that MIT report came out, several articles were published. A woman from Alaska said that the report was nonsense, that the women at MIT were just whining, and that they didn't have any hard data. That person's perspective was picked up by the media even though she didn't have any data either. This is a clear example of the problem with backlash. This one comment was given credibility, even though there was little to back it up. It was suggested that frequently there may be a backlash, so one should expect it and ignore it.

There is also the question of speed. How rapidly do organizations change? And how rapidly must they change? The answer to the first question is, Very slowly. How rapidly must they change? Much faster. Of course, those are the easy answers to the question. We discussed this in detail. Some people in our group reported that their institutions were working on this, that schools were doing equity surveys, that committees had already been established, and that administrations were listening. One person said that at her institution, she had seen real changes in 18 months, because all these reports are coming in and people are paying attention.

But you need somebody in a leadership position who is willing to pay the price. It may be a dean who is willing to say "Yes, we're going to do such and such, we're going to make these changes." Often these people pay a price for their actions. You need somebody who is willing to pay that price, but those people don't always exist.

There may be dangers in making changes too quickly. Sometimes if you promote a woman—or if you move her up into positions for which she doesn't have the proper training, background, or reputation—more damage is done than good. What happens is, everybody points to her and says, "Look, she was promoted without the proper credentials, without the proper background; she only got it because she is a woman." So the implication is that women who are not competent get promoted. Therefore, we have to be careful to make sure that the women who are promoted are properly trained for those positions. The counterargument to that, however, is that we really need to get our foot in the door. So there is some balancing to be done here.

We also heard that if we can move more quickly, if some data are collected and published on the lack of diversity in the faculty and so on, then alumni of these institutions would react by refusing to give money. This may be a mechanism we can use for quicker change. We really don't have that much time. I heard from a number of people that apparently the pipeline to that middle level is empty. While the situation varies from institution to institution, there is a feeling of urgency.

What are my conclusions? We need surveys. Surveys are important; they provide data, they provide ammunition. But we also need publicity. Publicity is very important, and it is especially important if we use it properly and for positive change.

Last but not least, Why is the makeup of this particular group—the workshop participants—what it is? Where are the deans, where are the department chairmen, where are the decision makers? A number of people said they had been sent by their dean or supervisors. But the supervisors are the ones who must take care of this problem. The question is, Why aren't those people here?

Lou Ann Heimbrook, Lucent Technologies: What I am going to be discussing today is plans for change. We have heard a lot about a variety of areas, and our group talked about where we need to go. I will share with you some points that you will be able to think about as you leave the workshop.

Let me start with plans for change. There is a need for linkage among industry, academia, and government. The first link we need is with the Morella Commission, which is an 11-person commission on the advancement of technology.[2] You should know about that commission because it is an avenue that will help you to follow up on some of the discussions we have had here. And you should link up with the AXXS '99 as well.

My second point concerns recommendations to the ACS Board of Directors and its Committee on Professional and Member Relations. We are all ACS members: How is our ACS board being run? Who are its members? What do we want them to be doing? What are they looking into? Are the issues raised here being looked at by the ACS Board?

Then there is political activity in your home states. Be politically active in supporting and sustaining the appointments or election of women to high positions at the federal, state, and local levels. Why is this important? If we look at the current Administration in Washington, we see women in various positions of power. We need to continually support this type of political activity, especially if we are going to be raising issues of women in the work environment, in this case, the field of chemistry. So we want people to know first that we applaud the efforts of women and second that we want them to

[2]Commission on the Advancement of Women and Minorities in Science, Engineering, and Technology Development (CAWMSET); the full Commission report, "Land of Plenty: Diversity As America's Competitive Edge in Science, Engineering and Technology," was published in August 2000.

continue. Whether that means being active at the federal, local, or state level, it is important to make sure that we keep up the activity.

The other thing is action—whatever is possible for us to do—to get an executive order applying Title IX to academic institutions. This is a tough one, a tough one. But we need to see what actions we can take, and clearly there is some related activity.

All you have to do is look around the United States and at the processes of government and you see that lobbying is one of them. Actively lobby your state and local legislators to support the advancement of women. In particular, push your institution to make this a priority. As you know, in industry, lobbying is an activity, and it can be for universities too.

We can work to implement a diversity ranking. Here are some ways of doing that. Use the ACS as a locus. Publicize the ranking. If it gets into print, let's make sure we use it well. Also, suggest and show the implications for the budget if enrollment drops because you do not fare well on this ranking.

This is an individual commitment; we have spent 2 days away from our respective industrial, academic, or governmental institutions. Commit to one action. What I plan to do is a series of actions. I think everyone is going to be taking action following this workshop. I am currently working in industry, so I will make sure that when my company looks at co-ops and at postdocs, we pay attention to the institution they are from. What is the male-female faculty ratio? What are its policies? Is the institution forward looking?

We also need to look at whether industry is giving back. I have probably not been involved in enough lecture activities. I need to get involved in that again, so people can see what I am doing.

We need to prepare and submit proposals to funding agencies on the issues that we have talked about here (an example is COACh). Work the issues into your proposals, look for funding for them, because it is only through money that we can truly address the problems.

We also need to engage our ACS student affiliates. This is very, very important. Students are our future, we know that.

We have heard that in industry we celebrate diversity. But industry is not yet the shining star it should be. There are a lot of faults, and we can still make many improvements.

Let's do just one activity: let's celebrate Women's History Month. That can be quite easy to do. It can be done in many ways, and one of them is celebrate your great scientists. As I look out at the audience, I see a lot of people I can invite to Lucent Technologies' Women's History Month. So you may be hearing from me.

Also, put these activities on the list of the ACS speakers bureau. This will help in getting industrial representatives to talk about these issues at universities. If visitors are speaking on their science, they can address these women's issues afterwards, in meetings with administrators. So, after you finish your excellent scientific talk, when they ask if you want to meet the dean, you should answer yes. Then you can bring up the issues we've been talking about here.

Dr. Hopkins is the one we clearly need to thank, not just for her excellent science but also for courage. We could use her as a template for how to become an agent of change in our institutions. If you are not at MIT but at another university where things are working well, then you have a template you can share with others.

The other thing that I would end with is this: Include women's issues in the purview of the ACS Office of Graduate Education. I hope this will stimulate some activity as we try to bring about change.

Nancy B. Jackson, Sandia National Laboratoriess: I was in that group, and I just wanted to add to one of your comments about the ACS Board of Directors. Since ACS has all the data, it could very quickly do a ranking of universities. So it would be possible to request—through one of the board members, several of whom are here—that ACS get these statistics in order and do a ranking ASAP. It may take some work to pull the data from the Directory for Graduate Research, but the data are there: they are collected by the ACS Committee on Professional Training. This is something that could be done and published quickly. I would like to see it in the student affiliate newsletter.

Robert L. Lichter, Dreyfus Foundation: I want to follow up on the comment about money. You don't need money to do a lot of things, but it surely does help to have it. I want to make an offer here: the Dreyfus Foundation will be very open to proposals that deal with many of these issues. I refer you to <www.dreyfus.org>. Look under the special grant program and follow the instructions.

Barbara K. Warren, Union Carbide: Is anybody here willing to set up a listserve so that we may share our ideas? ACS/CPT not only collects data on women faculty, but looks at how many women are on the faculty and what their level of development is. If it seems to ACS/CPT that women are not being promoted or are not getting sabbaticals or equipment, it asks questions. This is not done for every school, but it is something that ACS has been doing for a long time.

Participant: I think year 2000 is again the year for surveying all women chemists, am I not right? So when you get that survey form, please answer it, buy the book, and use it when you go to your dean, or your supervisor, or your director.

Shannon Davis, Solutia, Inc.: I want to take Barbara up on her challenge. I have hosted a Web site for quite some time for my own personal network, and now I am going to open it up. It is on yahoo.com, and I'll send everyone the address.

Janet G. Osteryoung, National Science Foundation: I would like to make a few concluding remarks. One of them has already been made both directly and indirectly by others, but I want to reinforce it. There are a lot of people here from senior ACS leadership. Your presence is very noticeable, and we really appreciate it. I think there is a potential for doing some good things.

Specifically, Frankie Wood-Black and I have already talked about having a symposium at the San Diego ACS meeting that is based on this workshop. We will work to develop that.

I would also like to thank the ACS for its assistance in getting Rep. Johnson to be dinner speaker last night. This was something that I don't think we could have done by ourselves.

Let me finally say something about NSF. Everything that has been said about what NSF does or doesn't do or thinks or doesn't think is very much in the folklore category. You have to be careful about things like that. It is a diverse organization that has a lot of temporary people moving through it, and one person's experience at one time in one part of the institution does not tell you very much about what NSF does or doesn't do. The practices and attitudes are actually quite variable.

In the Chemistry Division, I would say that we have worked diligently, although not very successfully, to do some positive things about the kinds of problems that have been discussed at this workshop. My mind has been stimulated to come up with some new ideas.

I should also mention that there is a new program in the works at NSF. Its clever name is Advance, and it will be designed to address some of the issues that were brought up here.

Finally, let me tell you two things people have said to me. One is, "I've changed my thinking." I think that is excellent. That is what you would like a workshop to do. The other is, "My head hurts!" I like that one a lot.

Appendixes

Appendix A

Workshop Participants

Richard C. Alkire, University of Illinois at Urbana-Champaign
Particia A. Andreozzi, The Dow Chemical Company
Michael W. Babich, Florida Institute of Technology
Susan J. Babinec, The Dow Chemical Company
Patricia A. Baisden, Lawrence Livermore National Laboratory
E. Kent Barefield, Georgia Institute of Technology
Marjam B. Behar, National Institutes of Health
Maryka H. Bhattacharyya, Argonne National Laboratory
Arthur Bienenstock, Office of Science and Technology Policy
Henry N. Blout III, National Science Foundation
Kristie A. Boering, University of California, Berkeley
Kristin Bowman-James, University of Kansas
Mairin B. Brennan, American Chemical Society
Maria K. Burka, National Science Foundation
Carol J. Burns, Los Alamos National Laboratory
Daryle H. Busch, University of Kansas
Allison A. Campbell, Pacific Northwest National Laboratory
Maureen G. Chan, American Chemical Society Board of Directors
Sally Chapman, Barnard College
Barbara J.W. Cole, University of Maine
Geraldine V. Cox, Eurotech, Ltd.
Naresh Dalal, Florida State University
Marcetta Y. Darensbourg, Texas A&M University
Tammy A. Davidson, East Tennessee State University
L. Shannon Davis, Solutia, Inc.
Donna J. Dean, National Institutes of Health

Stacy M. Dedinas, DuPont Engineering
Brian Doherty, American Chemical Society
Bernadette Donovan-Merkert, University of North Carolina, Charlotte
Ellen R. Fisher, Colorado State University
Suzanne E. Franks, Kansas State University
Victoria Friedensen, National Academy of Engineering
Hilary Arnold Godwin, Northwestern University
Christine S. Grant, North Carolina State University
Kimberly A. Gray, Northwestern University
Mary Lynn Grayeski, Research Corporation
Sandra C. Greer, University of Maryland
Richard M. Gross, The Dow Chemical Company
Jong-on Hahm, National Research Council
J. Howard Hargis, Auburn University
Lou Ann Heimbrook, Lucent Technologies
Janice M. Hicks, National Science Foundation
Nancy H. Hopkins, Massachusetts Institute of Technology
Sally S. Hunnicutt, Virginia Commonwealth University
Alison Hyslop, St. John's University
Nancy B. Jackson, Sandia National Laboratory
P. Wyn Jennings, National Science Foundation
Mary Welsh Jordan, American Chemical Society
Timothy A. Keiderling, University of Illinois at Chicago
Douglas A. Kiserow, U.S. Army Research Office
Valerie J. Kuck, Bell Laboratories
Anne Leslie, ARK Enterprises, Inc.
Robert L. Lichter, Camille & Henry Dreyfus Foundation
Andrew J. Lovinger, National Science Foundation
Mary L. Mandich, Bell Laboratories
Robert S. Marianelli, Office of Science and Technology Policy
Pamela Marino, National Institutes of Health
Gretchen Matthern, National Academy of Engineering
Linda B. McGown, Duke University
Claude F. Meares, University of California, Davis
Elise G. Megehee, St. John's University
Margaret V. Merritt, Wellesley College
Faith A. Morrison, Michigan Technological University
Sister Mary Murphy, St. Joseph College
Pushpalatha Murthy, Michigan Technological University
E. Ann Nalley, Cameron University
Lily M. Ng, Cleveland State University
Janet G. Osteryoung, National Science Foundation
Nancy L. Parenteau, Organogenesis, Inc.
Peter W. Rabideau, Iowa State University
Michelle Rice, DuPont
Geraldine L. Richmond, University of Oregon

Michael E. Rogers, National Institutes of Health
Celestine M. Rohlfing, National Science Foundation
Debra R. Rolison, Naval Research Laboratory
Nina M. Roscher, American University
Margaret W. Rossiter, Cornell University
Norka Ruiz Bravo, National Institutes of Health
John M. Schwab, National Institutes of Health
Cecily C. Selby, Radcliffe Institute for Advanced Study, Harvard University
Hratch G. Semerjian, National Institutes of Standards and Technology
Kathleen E. Sendall, Petro-Canada
W. Sue Shafer, University of California, San Francisco
Joan E. Shields, Long Island University
Jean M. Smolen, St. Joseph's University
Marylee Southard, University of Kansas
Maria Spinu, DuPont
Rosemarie Szostak, U.S. Department of Army
Elizabeth C. Theil, Children's Hospital Oakland Research Institute
Marion C. Thurnauer, Argonne National Laboratory
Laura Tubbs, Rochester Institute of Technology
Linda Tunstad, California State University
Virginia Valian, Hunter College
Isiah M. Warner, Louisiana State University
Barbara K. Warren, Union Carbide Corporation
Patricia L. Watson, DuPont Central Research and Development
Ming-Ying Wei, National Aeronautics and Space Administration
Kenton H. Whitmire, Rice University
Frankie K. Wood-Black, Phillips Petroleum
Catherine Woytowicz, American Chemical Society
Kristin Zimmerman, National Academy of Engineering

Staff
Maria P. Jones
Ruth McDiarmid
Susan R. Morrissey
Sybil A. Paige
Douglas J. Raber

Appendix B

Biographical Sketches of Workshop Speakers

Arthur Bienenstock is associate director for science of the Office and Science and Technology Policy (OSTP), the White House office that has responsibility for ensuring that the United States continues to maintain global leadership in science, mathematics, and engineering research and that science continues to provide support for the successful resolution of important problems in the areas of health, agriculture, the economy, energy, social well-being, education, and national security. Its science division concentrates on policy and interagency coordination directly related to the health of U.S. basic science, as well as on other policy matters that can be informed by basic science. Dr. Bienenstock received B.S. (1955) and M.S. (1957) degrees from the Polytechnic Institute of Brooklyn. He received his Ph.D. from Harvard University in 1962. In addition, he was a recipient of a Ph.D. (honorary) from Polytechnic University in 1997. He is the first recipient of the Pittsburgh Diffraction Society's Sidhu Award for his work in x-ray diffraction and crystallography. He is also a fellow of the American Physical Society and the American Association for the Advancement of Science.

At OSTP, Dr. Bienenstock has sought to gain general recognition of the interdependencies of the sciences and the need for the country to maintain broad scientific and technological strength. He has also focused on ensuring that the United States has a scientific and technological workforce, at all levels, that meets the nation's 21st century needs. Mindful of anticipated demographic changes, he initiated an interagency working group that is seeking to increase the participation of minorities, women, and the disabled in science and technology. He has led a task force on the government-university research partnership aimed at strengthening the relationship and has championed an interagency educational research initiative to fund large-scale, interdisciplinary research on teaching and learning.

L. Shannon Davis is the director of Industrial Products Technology at Solutia, Inc., a specialty chemicals company. She received her B.S. from Georgia Southern College and her Ph.D. in inorganic chemistry from the University of Florida. She began her career in 1988 at Monsanto's Pensacola, Florida, site as a senior chemist in the nylon intermediates business unit, where she worked on processes for the manufacture of adipic acid and hexamethylenediamine and to increase the value of coproduct streams from these processes and had responsibility for the on-site chemical pilot plant facility. In 1994, she was promoted to manager, product technology, in the Saflex business and moved to Springfield, Massachusetts, where she created and participated in a leadership team focusing on product and process improvements. After $2^1/_2$ years in Saflex, she returned to Pensacola as the manager of an R&D group responsible for technology supporting the carpet business. In 1997, Dr. Davis moved to her current position at corporate headquarters in St. Louis, Missouri. Her responsibilities include growth programs and existing technologies for heat transfer fluids, aviation fluids, metalworking fluids, and L-aspartic acid.

Nancy H. Hopkins is the Amgen, Inc., Professor of Molecular and Development Biology at the Massachusetts Institute of Technology. She obtained a B.A. from Radcliffe College in 1964 and a Ph.D. from the Department of Molecular Biology and Biochemistry at Harvard University in 1971. Her Ph.D. thesis, carried out in the laboratory of Mark Ptashne, dealt with gene expression in the bacterial virus lambda. Her postdoctoral research, under James D. Watson at Harvard and at the Cold Spring Harbor Laboratory, involved DNA tumor viruses. In 1973, she joined the MIT faculty as an assistant professor in the Center for Cancer Research, within the Biology Department, where she worked on mechanisms of replication and leukemogenesis by RNA tumor viruses for 17 years. She was promoted to associate professor in 1976, tenured in 1979, and promoted to professor in 1982. Nine years ago, Dr. Hopkins switched fields to work in developmental biology. Her laboratory first developed techniques for making transgenic zebrafish and is now using these techniques to isolate a significant fraction of the genes required for the normal development of the zebrafish embryo.

Dr. Hopkins is the author of numerous scientific papers in the fields of bacterial and animal viruses and in developmental biology, and she wrote, with four others, the fourth edition of the textbook *The Molecular Biology of the Gene.* She codeveloped and taught the first freshman biology course required of all MIT undergraduates, for which she was named a Class of 1960 Fellow. She is a fellow of the American Academy of Arts and Sciences (AAAS) and a member of the Institute of Medicine. In 1995 she was appointed chair of the first Committee on Women Faculty in the School of Science at MIT.

Debra R. Rolison is head of Advanced Electrochemical Materials at the Naval Research Laboratory (NRL). She received a B.S. in chemistry from Florida Atlantic University in 1975 and a Ph.D. in chemistry from the University of North Carolina at Chapel Hill in 1980 under the direction of Royce W. Murray. Dr. Rolison joined the Naval Research Laboratory as a research chemist in 1980. Her research at NRL focuses on the influence of nanoscale domains on electron- and charge-transfer reactions, with special emphasis on the surface and materials science of aerogels, electrocatalysts, and zeolites. Her program creates new nanostructured materials and composites for catalytic chemistries, energy storage and conversion (fuel cells, supercapacitors, batteries, thermoelectric devices), and sensors.

Dr. Rolison is a member of the American Chemical Society, AAAS, the International Zeolite Association, the Materials Research Society, and the Society for Electroanalytical Chemistry (SEAC). She wrote *Ultramicroelectrodes*, the first textbook in this very active research area of electrochemistry, with Martin Fleischmann, Stanley Pons, and Peter Schmidt. She and Henry White guest-edited an issue of *Langmuir* devoted to the electrochemistry of nanostructured materials (February 1999). Dr. Rolison was a member of the Advisory Board for Analytical Chemistry and is a current member of the editorial boards of the *Journal of Electroanalytical Chemistry* and *Langmuir*. She is a member of the Board of Directors for the SEAC and has served since 1997 as editor of the society's newsletter, *SEAC Communications.*

Margaret W. Rossiter is the Marie Underhill Noll Professor of the History of Science at Cornell University. She is the editor of the two official journals of the History of Science Society, the quarterly *Isis* and the annual *Osiris,* and the author of three award-winning books: *Justus Liebig and the Americans: The Emergence of Agricultural Science* (Yale University Press, 1975); *Women Scientists in America: Struggles and Strategies to 1940* (Johns Hopkins University Press, 1982); and *Women Scientists in America: Before Affirmative Action, 1940-1972* (Johns Hopkins University Press, 1995). She has also edited or coedited three other volumes.

Dr. Rossiter received her Ph.D. from Yale University School of Medicine in the history of medicine and science. She has held a number of fellowships, including a Guggenheim, a Rockefeller, and—from 1989 to 1994—a MacArthur, and has been honored with numerous teaching and research awards, including the 1997 Pfizer Prize and the History of Science Society award for the best book in English on the history of science for the previous 3 years. In the early 1980s she was on the first NSF Director's Advisory Committee on Equal Opportunities in Science and Technology and also served as program director for NSF's program in the history and philosophy of science. In 1996 and 1997 she was a fellow at the Institute for Advanced Study, where she worked on the history of science in the United States.

Kathleen E. Sendall is vice president of Western Canada Development and Operations at Petro-Canada. She is responsible for the Western Canada natural gas business and conventional oil properties, including technology and marketing functions that support the business. Ms. Sendall graduated from Queen's University, Kingston, with a B.Sc., honours, in mechanical engineering in 1977 and from the Western Executive Program of the University of Western Ontario Business School in 1990. Ms. Sendall joined Petro-Canada in 1978 after 2 years as an engineering supervisor of offshore and international joint ventures at Nova Gas Transmission. Ms. Sendall held various supervisory positions until 1991, when she was appointed wholesale marketing manager for Petro-Canada Products. In 1994, she served as manager and, later, director of business development in the Natural Gas Liquids business unit. She was appointed to her current position in 1996.

Ms. Sendall chairs the Advisory Council to the NSERC/Petro-Canada Chair for the Advancement of Women in Science and Engineering, Prairie Region. She is also a director of Junior Achievement of Southern Alberta and the Canadian Engineering Memorial Foundation of Calgary. She is a member of the Association of Professional Engineers, Geologists and Geophysicists of Alberta. In 1998, Ms. Sendall was the recipient of the YWCA Women of Distinction Award for the category "Business, Labour, the Professions and Entrepreneurs" and in 1999 became a fellow of the Canadian Academy of Engineering.

Virginia Valian has served as professor of psychology at Hunter College–CUNY since 1987. She has also held the position of director of the Hunter College master's program in psychology since 1991. She received her M.A. (1968) and Ph.D. (1971) in psychology from Northeastern University and her B.A. from the University of Michigan in 1963. She has held a variety of visiting faculty positions at universities in the United States and abroad, including Centre National de Recherche Scientifique (Paris), the University of Rochester, Wellesley College, Fudan University (Shanghai), Columbia University, and Mount Holyoke College.

Dr. Valian's research interests include first and second language acquisition, artificial language learning, and human sex differences. She is a member of the American Psychological Association, the American Psychological Society, and the Society for Research in Child Development. She serves as a member of the editorial board for *Cognition.* She has written and lectured widely and has received considerable acclaim for her 1998 book *Why So Slow? The Advancement of Women.*

Appendix C

Origin of and Information on the Chemical Sciences Roundtable

In April 1994, the American Chemical Society (ACS) held an Interactive Presidential Colloquium entitled "Shaping the Future: The Chemical Research Environment in the Next Century."[1] The report from this colloquium identified several objectives, including the need to ensure communication on key issues among government, industry, and university representatives. The rapidly changing environment in the United States for science and technology has created a number of stresses on the chemical enterprise. The stresses are particularly important with regard to the chemical industry, which is a major segment of U.S. industry, makes a strong, positive contribution to the U.S. balance of trade, and provides major employment opportunities for a technical work force. A neutral and credible forum for communication among all segments of the enterprise could enhance the future well-being of chemical science and technology.

After the report was issued, a formal request for such a roundtable activity was transmitted to Dr. Bruce M. Alberts, chairman of the National Research Council (NRC), by the Federal Interagency Chemistry Representatives, an informal organization of representatives from the various federal agencies that support chemical research. As part of the NRC, the Board on Chemical Sciences and Technology (BCST) can provide an intellectual focus on issues and fundamentals of science and technology across the broad fields of chemistry and chemical engineering. In the winter of 1996, Dr. Alberts asked BCST to establish the Chemical Sciences Roundtable to provide a mechanism for initiating and maintaining the dialogue envisioned in the ACS report.

The mission of the Chemical Sciences Roundtable is to provide a science-oriented, apolitical forum to enhance understanding of the critical issues in chemical science and technology affecting the govern-

[1]American Chemical Society, *Shaping the Future: The Chemical Research Environment in the Next Century,* Report from the Interactive Presidential Colloquium, April 7-9, 1994, Washington, D.C.

ment, industrial, and academic sectors. To support this mission, the Chemical Sciences Roundtable will do the following:

- Identify topics of importance to the chemical science and technology community by holding periodic discussions and presentations and gathering input from the broadest possible set of constituencies involved in chemical science and technology.
- Organize workshops and symposia and publish reports on topics important to the continuing health and advancement of chemical science and technology.
- Disseminate the information and knowledge gained in the workshops and reports to the chemical science and technology community through discussions with, presentations to, and engagement of other forums and organizations.
- Bring topics deserving further, in-depth study to the attention of the NRC's Board on Chemical Sciences and Technology. The roundtable itself will not attempt to resolve the issues and problems that it identifies—it will make no recommendations nor provide any specific guidance. Rather, the goal of the roundtable is to ensure a full and meaningful discussion of the identified topics so that the participants in the workshops and the community as a whole can determine the best courses of action.